# Springer Series on Biofilms

For further volumes:
http://www.springer.com/series/7142

Garth D. Ehrlich • Patrick J. DeMeo •
J. William Costerton • Heinz Winkler
Editors

# Culture Negative Orthopedic Biofilm Infections

Volume 7

*Editors*
Garth D. Ehrlich and J. William Costerton
Allegheny-Singer Research Institute
Center for Genomic Sciences Biofilm Research
Pittsburgh
Pennsylvania
U.S.A.

Patrick J. DeMeo
Allegheny General Hospital
Department of Orthopedic Surgery
Pittsburgh
Pennsylvania
U.S.A.

Heinz Winkler
Privatklinik Döbling GmbH
Wien
Austria

ISSN 1863-9607 ISSN 1863-9615 (electronic)
ISBN 978-3-642-29553-9 ISBN 978-3-642-29554-6 (eBook)
DOI 10.1007/978-3-642-29554-6
Springer Heidelberg New York Dordrecht London

Library of Congress Control Number: 2012954014



Printed on acid-free paper

Springer is part of Springer Science+Business Media (www.springer.com)

# In Memoriam

This book, like the entire Springer Biofilm series of books, was conceived, developed, and largely assembled by our dear friend and mentor in all things biofilm, J. William "Bill" Costerton. Unfortunately, during the final editing of this volume, Bill passed away from complications of pancreatic cancer, but true to his character and nature was gently guiding us through its completion even as he fought his last battle with enormous grace and dignity while surrounded by his large and loving family. Bill leaves behind an incredible body of work and is arguably the most important microbiologist of the past half century. It is not just his 700+ publications or his listing by ICI as one of the most cited microbiologists of all time that we make this claim. Rather it was his simple, but profound, realization that bacteria prefer to live in highly structured communities, which he termed biofilms, which set the stage for multiple revolutions across myriad fields of microbiology and infectious diseases upon which we base this assertion. His advances were made in disciplines as diverse as remediation of acid mine drainage, to the finding that many, if not most, "sterile loosenings" of periprosthetic joints result from biofilm infections by nontraditional pathogens, to the use of nanobacterial biofilm plugs for enhanced recovery from old oil wells, and to our modern day understanding of cardiac valve infections. Bill was the ultimate "big picture" guy and was most in his element when synthesizing findings across vast traditional scientific boundaries. Few scientists in the modern world can claim to have wrought such profound change in areas as diverse as water purification, petroleum production, environmental remediation, orthopedics, infectious diseases, urology, cardiology, . . ., and the list goes on, while simultaneously being an accomplished alpinist, backcountry skier, and all-around outdoorsman. To say that Bill was robust and that he lived his life large is an understatement; his lifelong incandescent and positive energy is already the stuff of legend, but perhaps

Bill can best be understood by his behavior at international scientific gatherings where he would often eschew the other potentates present to be able to have time to meet with the upcoming generation of folks, as it was the youth that asked the best questions according to Bill—and Bill was eternally young at heart.

Pittsburgh, PA Garth D. Ehrlich

# Preface

Because orthopedic infections often involve inert fixation and prosthetic devices and almost always involve sequestra of dead bone that act as foreign bodies, they are predominantly caused by bacteria living in biofilm communities. The first clinical consequences of this communal lifestyle of the causative bacteria were that these infections were not cured by host defenses or by systemic antibiotic treatment, and orthopedic surgeons have quickly developed the very effective strategy of aggressive debridement followed by very high dose antibiotic therapy. But the biofilm paradigm is complex, and the change from the acute paradigm of planktonic infections is profound, and these elaborate communities of bacteria living as highly integrated societies had one more surprise hiding up their tiny sleeves!

As Patrick DeMeo's Department of Orthopedic Surgery began to cooperate with Garth Ehrlich's Center for Genomic Sciences, several individual cases were identified and published as case reports, in which bacterial biofilms could be seen in infected tissues, but traditional cultures were negative. The routine hospital lab that provides services to Allegheny Hospital is accredited, but it does not incubate specimens anaerobically or keep cultures for more than 5 days, so the rates of positive culture are not as high as the more advanced labs at the Mayo Clinic. However, this level of microbiological service is typical of that available to most orthopedic surgeons in the USA. It was abundantly clear that DNA- and RNA-based molecular methods would be needed to replace cultures in the detection and identification of bacteria in orthopedic infections, but, initially, the "water was muddied" by the use of single species PCR methods that were nonquantitative and hypersensitive. Bill Costerton came to Allegheny Hospital just as the new universal DNA-based Ibis methods were being refined in Garth Ehrlich's lab, and he began to work with colleagues in Orthopedics to compare cultures with Ibis data and to confirm the results using RNA-based FISH probes and deep 16S sequencing.

The first Ibis study of infected total joints (with Fen Hu, Sandeep Kathju, and Nick Sotereanos) and the second study of infected nonunions (with Dan and Greg Altman) indicated that the universal Ibis technique improved the detection of bacteria from $<25$ % (cultures) to $>80$ %. Because this technology and several parallel platforms can provide results in 6 h and also give data on antibiotic sensitivity, we judged that we should announce this dramatic improvement in diagnosis in a special meeting (May 2011) attended by 100 colleagues who would be instrumental in demanding and managing this transformation. We have incorporated several presentations from this meeting, in the form of chapters in this book, ranging from parallel studies of modern detection techniques (Thomsen, Kennedy, and Shirtliff) to statements of how the modern methods will impact clinical practice (O'Toole, Sotereanos, and Parvisi). Perhaps the culmination of the book is a special chapter by Heinz Winkler, of Vienna, whose one-stage revisions of infected prostheses inspired much admiration at the May meeting and whose faithful adhesion to the biofilm concept has served hundreds of patients as a gold standard in the treatment of orthopedic biofilm infections.

The difficulty of detecting and identifying biofilm bacteria by culture methods is, of course, not limited to orthopedic infections. Less than 1 % of the bacteria growing predominantly as biofilms in natural ecosystems can be recovered by culture methods. Culture methods have been so insensitive in the detection of bacteria in major biofilm infections, like otitis media and prostatitis, that the bacterial etiology of these infections has been questioned and the polymicrobial nature of mixed species infections (e.g., diabetic foot ulcers) has been ignored when cultures grew very few species. The demonstration that modern molecular methods can detect and identify all of the bacterial species present in affected tissues in Orthopedics may have two salutary outcomes in all medical areas. First, the etiology of many chronic disease conditions will be better understood when all of the power of modern molecular diagnostics is brought to bear on them. Thus, elusive connections between infections and clinical problems such as nonunions at fracture sites will be resolved one way or the other. Second, clinicians will know the species identity of the infecting bacteria and their antibiotic sensitivities, as they design surgical or medical interventions, without any limitations imposed by the biofilm mode-of-growth or by the reluctance of certain species to grow in culture. It should be seen as simple justice as it was the orthopedic surgeons, who first understood the basic biofilm paradigm and modified their practice accordingly, that they benefit from the best diagnostic technologies that modern microbiology can provide.

3 March 2012

Garth D. Ehrlich
J. William Costerton
Heinz Winkler

# Contents

# The Problem of Culture-Negative Infections

**G.D. Ehrlich, P.J. DeMeo, and J.W. Costerton**

**Abstract** Because modern medicine suffers increasingly from the "silo" phenomenon, in which each specialty ponders its problems in isolation, the gradual emergence of a generalized threat to millions of patients is thus poorly countered by the disconnected efforts of small teams that address the same theme without the recognition of common ground. The recent recognition that bacteria have reverted to their natural biofilm strategy (Costerton 2007) in attacking human hosts, in response to advances in immunization (vaccines) and therapy (antibiotics), has been perceived in a piecemeal fashion that is slowly spreading amongst the silos. We respond to medical threats in relation to the immediacy of the dangers to the patient, so the first reaction was to the phenomenal resistance of biofilm infections to antibiotics and to host defense mechanisms, and the past three decades have seen a series of tactical maneuvers involving surgical resection and high-dose antibiotic therapy. While medicine reacted to this serious threat of overt bacterial infections that were not prevented by vaccination, and that persisted in spite of seemingly suitable antibiotic therapy, another equally serious biofilm problem was emerging at the bottoms of several silos. Experienced clinicians in many specialties saw cases in which they were certain that bacteria were involved, because all of the classical signs of infection were present, but the gold standard of diagnosis (culture) was negative. Some of these cases involved medical devices (Khoury et al. 1992), others involved infections of compromised tissues (Hoiby 2002), but the overall fight, conducted in isolation in many silos, was to decide on the correct antibacterial

G.D. Ehrlich (✉) • J.W. Costerton
Center for Genomic Sciences, Allegheny-Singer Research Institute, 320 East North Avenue, Pittsburgh, PA 15212, USA
e-mail: gehrlich@wpahs.org

P.J. DeMeo
Department of Orthopedic Surgery, Allegheny General Hospital, Pittsburgh, PA, USA

G.D. Ehrlich et al. (eds.), *Culture Negative Orthopedic Biofilm Infections*,
Springer Series on Biofilms 7, DOI 10.1007/978-3-642-29554-6_1,

strategy when the bacterial etiology of many important diseases (otitis media, prostatitis) was called into question by negative cultures. Bacteria do not respect the silos created by clinicians and scientists. They have switched from an acute frontal attack by planktonic cells, to a strategy of biofilm growth and chronic attack on infected tissues, and the most serious long-term effect of this tactical change may be that they evade detection by the classic methods of Medical Microbiology.

## 1 Silos in Clinical Medicine

It may be instructive to examine one particular medical silo (Ear, Nose and Throat = ENT) because the biofilm concept and the most refined molecular diagnostic capability came together in the team of Chris Post and Garth Ehrlich in that specialty. Culture data from otitis media with effusion (OME) were so consistently negative that some practitioners had suggested that it was a nonspecific inflammatory condition, and the basic bacterial etiology of the disease was cast into doubt. Clusters of bacterial cells could be seen by direct microscopy in the effusions from the ears of these patients, and DNA-based PCR methods (Post et al. 1995) showed the presence of very large amounts of DNA from the three major putative pathogens that occasionally grew in culture. When questions were raised about the viability of the bacteria in the effusions from the middle ear, the team used reverse transcriptase (RT)- PCR to detect short-lived messenger RNA (Rayner et al. 1998) to prove that the bacterial pathogens were both present and alive at the time of sampling, which allayed suspicions that antibiotic therapy alone could account for the negative cultures. In an elegant "coup de grace," the team then provided direct microscopic and molecular evidence (Hall-Stoodley et al. 2006) that OME is caused by bacteria growing in biofilms in the middle ear and that culture negativity is just as much a characteristic of this biofilm disease as is resistance to antibiotics and host defenses. In another silo (chronic wounds) the expert application of modern molecular techniques (Dowd et al. 2008) has proved that cultures only detect a small proportion of the bacterial and fungal species that are actually present, and that the clinical management of these complex infections can be radically improved using this accurate and pertinent information.

The inference from these scattered examples is disturbing because, if biofilm infections are indeed significantly more difficult to detect by conventional culture methods, tens of millions of patients are at risk for missed diagnoses. The CDC and the NIH have estimated that biofilm infections now constitute 65 and 80 % (respectively) of bacterial infections treated by physicians in the developed world, and a recent publication reports (Wolcott et al. 2010) that these infections affect 14 million, and kill > 400,000, Americans each year. Culture methods are the only FDA-approved and universally available technology for the detection and identification of bacterial and fungal infections in most of the developed world. In view of these well documented failure of cultures to detect such common biofilm infections as OME in children, and chronic UTI and prostatitis in adults, we must speculate concerning whole disease categories that are presumed to be nonbacterial because of consistently negative cultures. Perhaps peptic ulcers are not the only disease whose therapy will change radically when its etiology is more accurately understood, and perhaps more individual patients will be treated more effectively when the identity of any and all bacteria in their tissues can be accurately determined. We live in the era of precise DNA-based forensics, and of whole genome sequencing, and patients will be best served when we mobilize these resources for the prompt and accurate diagnoses of bacterial disease.

## 2 The Rational Basis of the Problem of Culture Negativity

Bacterial biofilms have now been studied very extensively, in the very well funded ($100 million to date) Center for Biofilm Engineering (CBE) and in hundreds of labs worldwide, and a clear picture of the structure and function of these very successful communities has emerged (Fig. 1). The majority of the bacterial cells in most microbial communities grow in the biofilm phenotype, and indulge in cooperative metabolic processes, as depicted in the area around the label "heterogeneity" in the middle of the figure. These cooperative processes are facilitated by physical connections between the individual cells like nanowires (Gorby et al. 2006), nanotubes (Remis et al. 2010), and temporary connection via pili and structured eDNA (Whitchurch et al. 2002; Goodman et al. 2011), and any attempt to rip one of these resolutely interactive cells from its designated bed in the biofilm would result in a group of "dazed and confused" bacterial cells that would be unable to grow in any medium. We have transferred well washed "chunks" of single species biofilms from flow cells in which they were growing, and they have failed to establish colonies on the surfaces of agar plates on which they were placed under direct microscopic observation. Pradeep Singh has recently shown (Singh P Copenhagen biofilm meeting 2011) that cells of *Pseudomonas aeruginosa* undergo a structural change in the lipopolysaccharide (LPS) of their outer membranes that makes them unable to grow on the surface of agar plates, even if the medium is otherwise ideal for the cultivation of Pseudomonads. Biofilm cells express a radically different phenotype (Sauer et al. 2002) from that of the planktonic cells that have been studied in the laboratory, for more than 150 years, and one of their salient characteristics is that they fail to grow and produce colonies when they are "ripped from their beds" in their "cozy communities" and placed on the surfaces of agar plates.

Virtually all of the cells in a mature biofilm, in vitro and during infection, operate in the biofilm phenotype, in which the pattern of gene expression may vary from the planktonic phenotype by as much as 70 % of the expressome, and this accounts for culture negativity in many cases. However, biofilms maintain a dispersal strategy to ensure that they can colonize distant sites in their ecosystems. This results in a reversal of the protective strategies in some locations of the community to produce

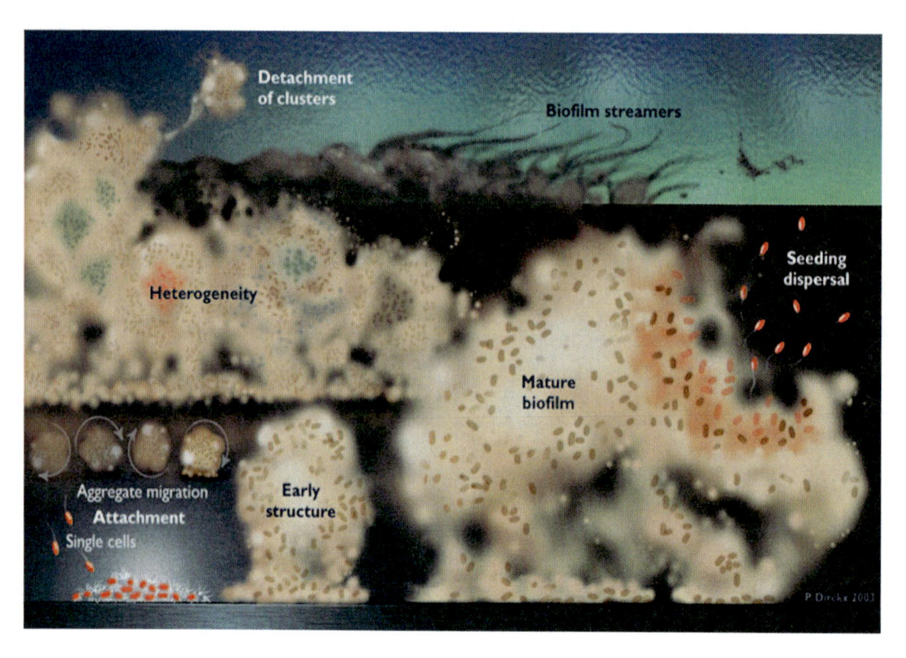


**Fig. 1** Diagrammatic representation of the location and roles of planktonic and biofilm cells in a typical biofilm community

planktonic cells in a programmed manner (Sauer et al. 2004) so that significant numbers of these dispersive cells are constantly produced (Fig. 1 at 3 o'clock).

In the grand Microbial Ecology design the planktonic phenotype of each species is designed to travel far from the biofilm community, often following favorable gradients, to "find a new home" and establish new communities. These streamlined planktonic cells have no cell–cell connections, their cell walls are adapted for stability in a variety of environments, and they are specially adapted to adhere to surfaces (including agar) and to form new communities, and it is these planktonic cells that culture methods were originally designed to detect. Before the development of vaccines, planktonic bacteria from an environmental source (e.g., cholera) could enter the human gut, overwhelm the local defenses, and set up a finite number of planktonic cell "factories" that would constitute a devastating infection. Diagnosis was simple, planktonic bacteria would grow on agar plates, and modern engineering was mobilized to contain the epidemic. Other epidemic diseases (e.g., typhoid and diphtheria) had human reservoirs, but the disease was still acute in that planktonic cells attacked the host and either overwhelmed its defenses in a week or less, or left the survivors immune and generally stronger for the encounter. Again, cultures gave a good indication of the presence of planktonic bacteria, and of the species that was present, and guided both preventative and therapeutic strategies.

When acute bacterial infections became much less common, because vaccines built up effective levels of immunity, and when the medical response to these diseases improved because of the development of antibiotics that killed the planktonic cells, the bacteria evoked their basic alternative strategy and began to attack as biofilms. These biofilm diseases, which were first noticed in connection with medical devices and with compromised human tissues (e.g., cystic fibrosis), initially caught our attention because they persisted in the face of active host defenses and because they are very profoundly resistant to conventional antibiotic therapy (Costerton et al. 1999). But another characteristic of biofilm infections soon began to emerge, which was that cultures were very ineffective in the detection and identification of their causative organisms. This culture-negative phenomenon is entirely logical, because cells in the biofilm phenotype grow so poorly on agar, and the problem was complicated in the case of biofilm infections, by the fact that antibiotics had often been used for extended periods of time so that the planktonic cells which might have indicated infection had been killed. These facts combined to produce a nadir in diagnoses, because the bacteria that actually cause biofilm infections are very difficult to grow in culture, and because the planktonic cells whose presence would indirectly indicate the presence of a biofilm had been killed by host defenses or by antibiotics. Various piecemeal strategies have been tried to improve detection by cultures, including the sonication of medical devices suspected of harboring biofilms (Trampuz et al. 2007), and medical laboratories with special mandates (e.g., Mayo Clinic) have explored the use of special media to grow pathogens that are recalcitrant to culture, but special culture methods travel poorly in the cash-strapped milieu of routine cultures. The problem of culture-negative biofilm infections remains complicated, it threatens the lives of millions of patients, and it constitutes an emergency because these infections now comprise the vast majority of bacterial infections and the only FDA-approved method for their detection has a sensitivity of approximately 20 % (Costerton et al. 2011).

## 3 Molecular Answers to the Culture-Negative Problem

Biofilm cells contain DNA in amounts approximately the same as planktonic bacterial cells and this DNA can be analyzed to determine the species identity of one or more organisms, and the presence of specific genes that cause antibiotic resistance. DNA analysis has gradually risen from a supportive role in forensics, to a legally sacrosanct basis for child support payments or release from custody for crimes of passion, so its accuracy must be said to be the "gold standard." DNA analysis forms the technical basis for the identification of all of the bacteria naturally associated with human tissues, in the new Human Microbiome Project and in virtually all studies of bacterial populations in natural ecosystems (Hugenholtz et al. 1998), but the pace of these studies is leisurely and may extend for weeks or months. The great DNA diagnostics machine whirs in mysterious ways, to the amazement of this humble morphologist, and the time necessary to detect specific cancer markers drops from weeks to days. The immediacy mandated by the need to detect bacterial bioterror agents has stimulated the development of a new DNA analysis method that delivers accurate data on the presence of any and all bacterial and fungal species in ±6 h. This method also detects the genes responsible for methicillin and vancomycin resistance in the same time frame. This technology (the Ibis PLEX-ID) will certainly not be the only or the definitive technology that will replace cultures in the diagnosis of bacterial infections, but the message is that DNA-based technologies have the speed to be useful in the diagnosis and management of biofilm infections.

The unblinking eye of the great DNA diagnostic machine will deliver quantitative data on the bacterial and fungal DNA present in the tissues it has analyzed, and physicians and surgeons will be projected back to the worst lectures in their least favorite class in medical school. If we walk through the process of molecular diagnosis, as it will become available during the next 2–5 years, we may be able to reduce the pain that comes when we move from seeing "through a glass darkly" to seeing all of the microbiology detail in stark bright light. If experienced surgeons simply ignore organisms they have never heard of, and concentrate on the familiar *Staphylococcus aureus* (especially MRSA), coaggulase-negative Staphylococci (CONS) and various Streptococci, we will see that these known pathogens are detected accurately and with much more sensitivity. The most pedantic amongst orthopedic surgeons may fuss about whether the very large number of MRSA in an infected prosthesis are alive or dead, which we cannot determine by the basic DNA technique, but the clinical decision to remove the prosthesis to debride extensively and to cover the wound with vancomycin can be taken with a high degree of confidence. In complex multispecies infections like diabetic foot ulcers, the detection of Candida species in a cohort of patients who had responded poorly to antibacterial antibiotics has triggered the decision to use ketoconazole, with excellent outcomes (Dowd et al. 2008). If clinicians simply stay within their comfort levels, treating organisms which they recognize as pathogens in their own specialties, and ignoring all reports of bacteria and fungi that they do not recognize, biofilm infections will be recognized earlier and treated more effectively. The universal Ibis technology that detects any and all bacterial DNAs is already "filtered" to remove any detections in which less than 3 of the 16 PCR primers did not "fire" to produce amplicons, and the system can be adjusted so that DNA present in very small amounts (e.g., <10 genomes/well) or DNA characteristic of known nonpathogens are simply not reported unless requested.

Physicians and surgeons who have treated biofilm infections during the era in which cultures have predominated have formed a bewildering variety of impressions of the comparative pathogenicity of various microorganisms. In a single coffee nucleus I have been told that *Staphylococcus epidermidis* is a nonpathogenic "contaminant" by one prominent surgeon, while another even more prominent surgeon said that "Staph epi" is a horrible pathogen probably worse than MRSA and that it will "eat the whole leg." We will need a prolonged period of accurate microbiology to resolve these issues. In orthopedic infections involving single organisms it will be instructive to assemble all clinical data to determine

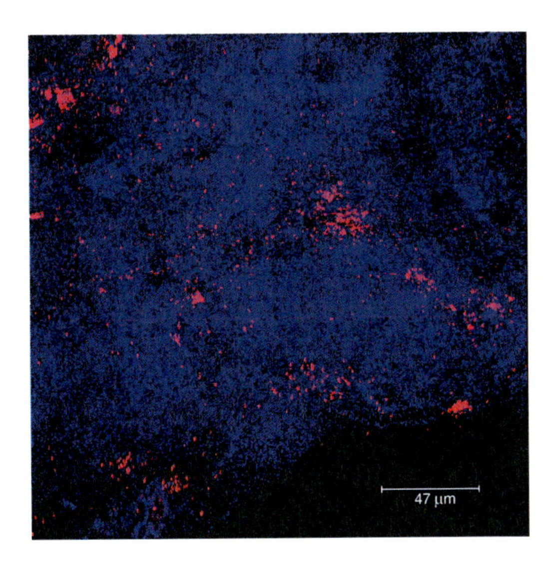


**Fig. 2** Confocal micrograph of tissue from a nonunion secondary to an open fracture of the tibia stained with the species-specific 16S FISH probe for *Enterococcus faecalis*

whether *Enterococcus faecalis* can actually cause the failure of a prosthesis or the nonunion of a fracture. When we visualize cells of *E. faecalis*, in tissues adjacent to a nonunion secondary to an open fracture (Fig. 2), we see that large areas of the tissue (blue in reflected light) are permeated by these coccoid cells, and that they have formed extensive biofilms in several areas. There is no possibility of error in these direct microscopic data, because human tissues contain none of the bacterial 16S rRNA against which the probe is directed, and because the bacterial cells we see are cocci less than 1 μm in diameter that are integrated within the tissue.

These direct microscopic data tell us that cells of *E. faecalis* had occupied tissues adjacent to this nonunion, and that they were present as biofilms, but only clinical data on a cohort of similar patients infected only with this organism will gradually allow us to assess the relative pathogenic potential of this organism in orthopedic infections. If it transpires that *E. faecalis* can cause overt orthopedic infections, alone or in combination with other species, it can be added to the list of known pathogens and suitable antibiotic therapy can be used in combination with surgical debridation.

*Propionobacterium acnes* is another bacterium whose role in orthopedic infections is sporadically invoked, even though it is a recognized pathogen in infections following neurosurgery (Nisbet et al. 2007). This gram-positive rod becomes predominant in the human skin (especially in sweat glands) just prior to sexual maturity, particularly in the axilla, and it is rarely detected by culture because it is slow-growing and requires special media and anaerobic incubation conditions. We have found *P. acnes*, often in very large and elaborate biofilms, in nonunions secondary to open fractures in which elements of the broken bone have penetrated the skin before the fracture was surgically reduced with internal fixation. However, the most extreme *P. acnes* biofilm (Fig. 3), we have seen in >200 orthopedic cases, was in the tissue removed from a painful locus adjacent to a healed fracture. The very

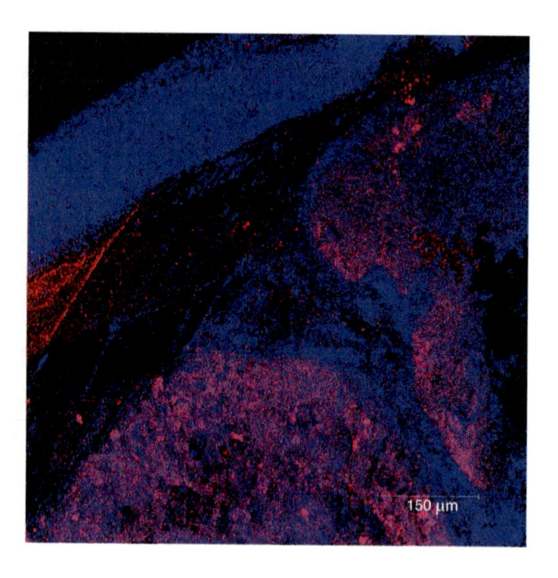


**Fig. 3** Confocal micrograph of tissue removed from a painful locus adjacent to a healed fracture, and reacted with the species-specific 16S FISH probe for *P. acnes*

extensive biofilm, which is pink where the bacteria are integrated into tissue, and red where the bacterial biofilm is growing between tissue elements, has not invaded all of the available tissues (see blue region at the top of the micrograph), and its presence did not prevent the healing of the fracture.

These direct observations of the structure of *P. acnes* biofilms will be replicated in cohorts of patients in which prosthesis failure or nonunion was associated with the presence of this organism alone, and clinical data will be mobilized to determine whether *P. acnes* alone can cause these orthopedic misadventures. In the event that this organism does proved to be a *bona fide* pathogen, therapy will be simplified, because this arcane organism is still sensitive to penicillin G and the neurological lesions were readily cured by debridation and penicillin therapy (Nisbet et al. 2007).

Sensitive methods for the detection of bacteria will put an end to the archaic concept that the human body is a largely sterile edifice, into which bacteria make occasional damaging intrusions. The presence of *Helicobacter* in the human stomach was not detected until special culture techniques were developed following the discovery of its role in duodenal ulcers (Forbes et al. 1994), bacteria have been found deep in the human female reproductive tract and some have been associated with normal pregnancy (Romero et al. 2008), and it is patently clear that many organisms leave their usual lairs in the human microbiome and circulate as transient bacteremias. Most replacements of knee and hip joints are to correct advanced osteoarthritis that has persisted for decades, and molecular methods have shown that ±30 % of the joints that are removed in primary arthroplasties show the presence of *T. denticola*, which is a spirochete whose pathological reservoir is the infected sulcus in periodontitis. We can see the spiral cells of *T. denticola* (Fig. 4) associated with the tissues of osteoarthritic joints that have been removed in primary arthroplasties, in samples stained with species-specific FISH probes, but we can only speculate concerning whether these organisms may play any important role in the arthritis.

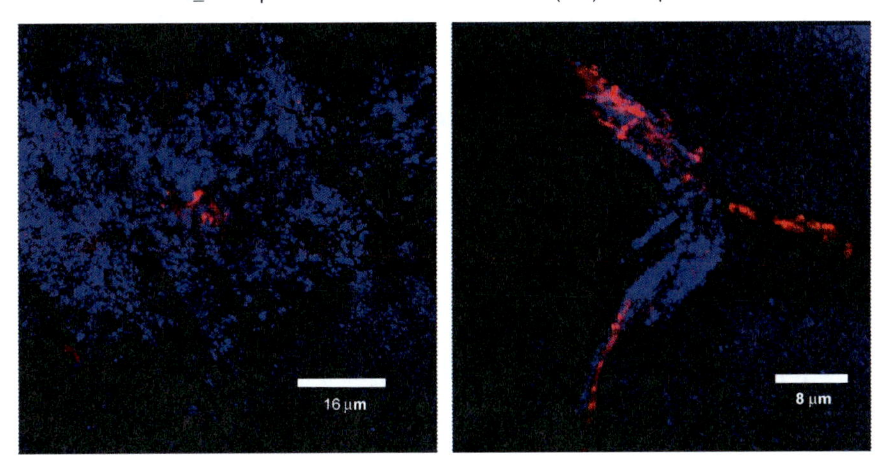


**Fig. 4** Confocal micrograph of a preparation of tissue from an osteoarthritic knee, which was replaced by a primary arthroplasty, which has been stained with the species-specific 16S FISH probe for *Treponema denticola*

It may be germane to note that the elongated spiral cells of *T. denticola* seen in these preparations are seen in complex biofilm-like structures that are adjacent to the human tissues (blue diffuse), and are rarely seen to penetrate these tissues. This spatial arrangement may reflect an association, in the absence of a causal relationship. Now that we can detect the presence of any and all bacteria in orthopedic samples, we will gather clinical data to determine whether the presence of this notably motile dental pathogen has any significance when, as often happens, it is present in a total joint infection (TJI).

# 4 Multiple Paradigm Shifts

While the whole world of bacterial population analysis moved from cultures to DNA-based methods in the early 1980s, Medical Microbiology was prevented from making this paradigm shift, because of the necessity for rapid results to guide life-and-death decisions. This paradigm shift from cultures to DNA-based methods coincided with another major paradigm shift in Medical Microbiology, in which acute diseases that are easily detected by culture methods were gradually replaced by biofilm diseases that are inherently difficult to detect by culture but are readily detected by DNA-based methods. The result of this double paradigm dilemma has been that physicians and surgeons who treat infected patients, and infectious disease specialists who consult in this life-saving process, have been faced with increasing numbers of patients who appear by all criteria to be affected by bacterial infections, but whose cultures are negative. A double paradigm shift puts the vulnerable human being at risk for "intellectual whiplash," but the speed of the novel DNA-based methods has now been improved such that they are now much more rapid than cultures, and the concept of biofilm infections is now widely accepted in most of the silos of modern medicine. The imminent commercialization of the Ibis system in the USA, and of the SeptiFast system in Europe, will best serve the medical community (and its patients) if we can inform the mental gymnastics necessary to accomplish this double paradigm shift in Medical Microbiology.

When a dim flickering light has barely illuminated the dangerous bacteria in each clinical area, the names of the really dangerous pathogens are well known, and the brighter light will show the same organisms and will reveal their antibiotic sensitivities in a more timely fashion. Notorious pathogens like MRSA will be detected rapidly, and the biofilm concept will be invoked to recommend that these protected bacterial communities be removed by careful surgery, and "mopped up" with specific antibiotics. This will be the practical baseline improvement provided by the double paradigm shift but, for the more perceptive, the increased sensitivity of the new molecular methods will report the presence of bacterial species whose role in pathogenicity is not yet known. Accurate microbiology can then be combined with careful clinical research to study cohorts of patients whose surgical samples contain only one putative pathogen to establish, unequivocally, that the presence of that organism can cause nonunion or implant failure. While dim lights are comfortable, and we gradually come to understand everything that we can see, wonderful things happen when we turn the light on to full intensity and see all of the bacteria that are arrayed against us: and maybe even some that are on our side!

# 5 Historical Origins of the Problem of Culturability

Antony van Leeuwenhoek used his primitive microscope, in 1716, to tell the world that "animalcules" less than 1 μm in size existed in the "scuff" from his non-too-immaculate teeth, and that they betrayed their viability by swimming vigorously through fluids. This curiosity went largely unremarked until, as is often the way in science, these small creatures were implicated as threats to humans survival and to the Gross National Product of France in the early to mid-1800s. The threat to the French GNP involved the death of silkworms and strange tastes that occurred in iconic wines and cheeses (mille horreurs!), and the white knight was Louis Pasteur. Louis was process-oriented, he was aware of the polymicrobial nature of most microbial ecosystems, and he bent pathogens and symbionts (alike) to his will by manipulating whole systems so that whole microbial communities obeyed human commands. The process of pasteurization and the burgeoning and vibrant Institute Pasteur, in Paris and in many other locations (including Tahiti), stand as tributes to his genius and his perceptions. Hundreds of processes, ranging from simple sewage treatment to the production of taxol by exquisitely engineered strains of microbes from the bark of the Pacific yew tree, can all trace their intellectual roots to Louis' concept of bacteria as members of integrated communities.

The response to the threat posed by bacteria to human health was equally successful but, in conceptual terms, the polar opposite. When faced with the problematical haystack of mixed species communities, Robert Koch chose to extract the "needle" and to examine it minutely, and in isolation from the neighbors and partners it would have had in the haystack. This separation of the needle from the haystack relied on a quirky but brilliant observation that certain fast-growing bacterial cells, especially those that caused acute epidemic infections, would grow to produce visible colonies when placed on the surface of an agar plate and separated from their fellows by progressive "streaking." The agar had to contain nutrients and salts similar to human blood or (in some cases) blood itself, the temperature and gas phases had to be optimal, and most important pathogens would grow to produce colonies with colors and shapes that betrayed their presence to the practiced eye. The arcane science of the detection and identification of bacteria by culture sprang up, with coffee Klatsch to exchange media and compare culture conditions, and mysterious additions like egg to encourage *Mycobacteria*, and Fisheria extract to tease *Legionella* out of the haystack and allow it to grow sufficiently to produce colonies. A prematurely wise graduate student in the CBE once observed that cultures resemble gardening, in that one "drags a rake" along the path of a mixed species English garden, then shakes the rake onto rich potting soil and observed (after a few weeks) what has grown! If a plant was reproducing at the time of sampling propagules would thrive and we would record its presence, but if it was not propagating or if its seeds did not "like" the potting soil, its presence would go unrecorded even if it dominated the ecosystem being studied. The skills necessary to recover bacteria by culture were largely experiential, and often found in people who had failed mathematics in high school (like one of us = J.W.C.), and cultures received a further fillip when the whole battery of methods to determine antibiotic sensitivity were added to the paradigm. The Koch Institut, located in a leafy suburb of Berlin, is the quintessential beautifully equipped laboratory in which a team of more than 200 brilliant microbiologists will certainly refine the DNA-based techniques that will replace the culture methods that enabled its founder to (virtually) wipe out epidemic bacterial disease and to save millions of lives.

The decision to remove bacteria from their integrated communities, by culture methods, was a very effective strategy at the time that it was adopted in the mid-1800s, and its utility has continued until the end of the twentieth century. Culture methods were successful in the detection and identification of bacterial pathogens, and the data that they provided enabled the largely complete eradication of epidemic diseases, even before the modern antibiotic era. In modern times, cultures have been pivotal in the discovery of antibiotics, and in the application of antibiotics to the virtual conquest of acute bacterial infections in uncompromised patients in the developed world. While there were no practical alternatives to culture methods, these venerable techniques served us well, and they still comprise the only FDA-approved methods for the recovery of bacteria from clinical specimens and for the determination of antibiotic sensitivity. Culture methods will, of course, continue to be the preferred methods of cultivating bacteria for study in the laboratory and for certain applications like checking strain purity, but

Microbiology is an evolving science and the utility of all widely used techniques should be evaluated by clinical microbiologists on a regular basis.

# 6 Two Solitudes

Many microbial processes are very important for humans and two fundamentally opposite approaches to their management developed, along parallel paths, for the last 150 years. Engineering is not tolerant of failure, so processes like wastewater treatment were managed by tweaking physical parameters, like flow and oxygen supply, until the microbial systems involved behaved properly and clean water came out of taps with monotonous regularity. If additional microbial activities were required, like the removal of phosphate, these same properties were tweaked in different ways until the whole system behaved correctly and the combination of factors that gave success was recorded and reproduced *ad nauseam*. Engineers do not like complexity or surprises, so they paid very little attention to the individual bacterial species that oxidized organic materials or bound phosphate, but they hit their performance windows by knowing how the whole ecosystem responded to physical and chemical variables. They were, without necessarily knowing it, Pasteurians. Essentially, they joined the pastoral people who, from time immemorial, have taken milk or mixtures of barley and hops and gently nudged the natural microbial ecosystem in the direction of perfection by subtle manipulations.

Another bunch of engineers joined their microbiological allies to tackle the problem of microbially influenced corrosion (MIC), which destroyed marine structures and metal pipelines at an alarming rate. Because microbial biofilms destroy metals by setting up classic "corrosion cells," in which they create cathodes by their metabolic activities and stimulate the formation of corresponding anodes from which metal is lost, engineers imposed cathodic protection currents on susceptible metals and solved the marine problem overnight. They did not culture or identify the bacterial species responsible for MIC, they did not even know the details of the metabolic activities that created the cathodes, but they found that the whole process within this damaging ecosystem could be halted by imposing an overriding DC potential and the problem was solved. Cathodic protection cannot prevent MIC in the interior lumen of pipelines, for complex physical reasons, so the same team of engineers and microbiologists used the whole system approach to detect and prevent corrosion in the millions of miles of pipes that interlace our world. They developed a test in which a steel nail is suspended in an anaerobic medium, and they mobilize their defenses when the nail and the medium turn black, when the nameless organisms set up corrosion cells. Then they regularly scrape the inner surfaces of the pipe with scrapers called "pigs," they discourage the dispersed microbial community with universal biocides, and they chalk up another victory for the Pasteurian approach.

The Kochian approach to Microbiology seems counterintuitive, in light of our current grasp of systems ecology, but it suits one human imperative, and it has saved (literally) hundreds of millions of lives. This approach, which is embodied in

Koch's principals, is perfectly suited for the management of situations in which one bacterial species causes damage to the human body or to an important human asset, by its own unique activities. The imperative is imposed by organisms that can survive in mixed species ecosystems and produce toxins and/or enzymes that kill or damage humans or human assets, and we need to only think of the strains of cholera or phyloxera that killed millions and significantly degraded the quality of life of many others. Other instances would include bacterial species that can invade normally sterile organs like the brain or the liver, by dint of special invasive mechanisms or by nosocomial routes, and whose very presence in these protected redoubts causes damage or death. When the agent of a particular damage is an individual bacterial species, the removal of the needle from the haystack yields benefits in accurate diagnosis and effective treatment, and represents the correct approach to the problem. Certain bacteria, such as *Escherichia coli O-157* and *Listeria monocytogenes*, should not be in our food and any culture or PCR method that detects them is valued highly.

The success of the Kochian approach has, paradoxically, limited its relevance in the modern world. Early in the last century the isolation of the viral and bacterial agents of acute epidemic disease facilitated the development of vaccines that converted immunologically naïve human populations to functional resistance, and allowed these threats to flicker out in the developed world. Other bacterial diseases that were readily identified by acute symptoms, and easily detected by classic culture techniques, were gradually brought under control by specific antibacterial agents that evolved from simple sulfa drugs to the most complex DNA gyrase inhibitors. In the past eight decades virtually any pathogen that had the temerity to kill or threaten humans or important human assets has been the subject of a concerted counter attack by the vaccine industry or by big pharma or big agriculture, and their elimination is not complete but is (in most cases) in progress. The specialized bacterial pathogens that represented specific needles that could be removed from the haystack by culture methods have been defined and studied and analyzed, and many of them are receding into oblivion and irrelevance. But we must be wary because even the most specialized of pathogens (e.g., *Vibrio cholera*) may retreat into the environment and emerge with full virulence, if we relax our vigilance and surveillance.

Modern microscopy, like the two-photon confocal microscope with coupled deconvolution software, has allowed us to visualize all of the bacteria on solid surfaces in natural environments and on tissue surfaces in the human body. The impact of this research was devastating to traditional microbiologists in that planktonic bacteria, which had been studied assiduously by microbiologists for 150 years, were seen to comprise $<1$ % of the bacterial cells in natural and pathogenic ecosystems. The vast majority of bacteria live in integrated biofilm communities (Fig. 1), and the only exceptions to this general rule are (perhaps ironically) human constructs like laboratory cultures and brewer's vats, and human and other animal bodies under attack by planktonic cells of specialized pathogens. This small Kochian cycle has now run its course, because of vaccines and antibiotics, and all of the microbial systems that influence human life are now seen in terms of Pasteurian principles in which bacteria are members of functional integrated communities which can be manipulated and controlled as whole communities. Planktonic systems are still of some minor interest, and the culture techniques that serve them well are still valid in that limited context, but virtually all of our microbial problems now involve biofilms, and we need to find room for the Petri plates somewhere between our quill pens and our hand-wound gramophones.

## References


Costerton JW (2007) The biofilm primer. Springer, Heidelberg, 200 p

Costerton JW, Stewart PS, Greenberg EP (1999) Bacterial biofilms: a common cause of persistent infections. Science 284:1318–1322

Costerton JW, Post JC, Ehrlich GD, Hu FZ, Kreft R, Nistico L, Kathju S, Stoodley P, Hall-Stoodley L, Maale G, James G, Shirtliff M, Sotereanos N, DeMeo P (2011) New rapid, and accurate methods for the detection of orthopeidic infections. FEMS Immunol Med Microbiol 61:133–140

Dowd SE, Wolcott RD, Sun Y, McKeehan T, Smith E, Rhoads D (2008) Polymicrobial nature of chronic diabetic foot ulcers using bacterial Tag encoded amplicon pyrosequencing (bTEFAD). PLoS One 3:e3326

Goodman SD, Obergfell KP, Jurcisek JA, Novotny LA, Downey JS, Ayala EA, Tjokro N, Li B, Justice SS, Bakaletz LO (2011) Biofilms can be dispersed by focusing the immune system on a common family of bacterial nucleiod associated proteins. Mucosal Immunol 4:625

Forbes GM, Glaser ME, Cullen DE, Collins BJ, Warren JR, Christiansen KJ, Mashall BJ (1994) Duodenal ulcer treated with Helicobacter pylori eradication: seven year follow-up. Lancet 343:258–260

Gorby YA, Yanina S, McLean JS, Russo KM, Moyles D, Dohnalkova A, Beveridge TJ, Chang IS, Kim BH, Kim KS, Culley DE, Reed SB, Romine MF, Saffarini DA, Hill EA, Shi L, Elias DA, Kennedy DW, Pinchuck G, Watanabe K, Iishi SI, Logan B, Nealson KH, Fredrickson JK (2006) Electrically conductive bacterial nanowires produced by Shewanella oneidensis strain MR-1 and other microorganisms. Proc Natl Acad Sci USA 103:11358–11363

Hall-Stoodley L, Hu FZ, Gieseke A, Nistico L, Nguyen D, Hayes J, Forbes M, Greenberg DP, Dice B, Burrows A, Wackym PA, Stoodley P, Post JC, Ehrlich GD, Kerschner JE (2006) Direct detection of bacterial biofilms on the middle ear mucosa of children with otitis media. J Am Med Assoc 296:202–211

Hoiby N (2002) Understanding bacterial biofilms in patients with cystic fibrosis: current and innovative approaches to potential therapies. J Cyst Fibros 1:249–254

Hugenholtz P, Goebel BM, Pace NR (1998) Impact of culture-independent studies on the emerging phylogenetic view of bacterial diversity. J Bacteriol 180:4765–4774

Khoury AE, Lam K, Ellis B, Costerton JW (1992) Prevention and control of bacterial infections associated with medical devices. ASAIO Trans 38(3):M174–M178

Nisbet M, Briggs S, Ellis-Pegler R, Thomas M, Holland D (2007) Propionobacterium acnes: an under-appreciated cause of post-neurosurgical infection. J Antimicrob Chemother 60:1097–1103

Post JC, Preston A, Aul JJ, Larkins-Pettigrew M, Ridquist-White J, Anderson KW, Wadowsky JM, Reagan DR, Walker ES, Kingsley LA, Ehrlich GD (1995) Molecular analysis of bacterial pathogens in otitis media with effusion. J Am Med Assoc 273:1598–1604

Rayner MG, Zhang Y, Gorry MC, Chen Y, Post JC, Ehrlich GD (1998) Evidence of bacterial metabolic activity in culture-negative otitis media with effusion. J Am Med Assoc 279:296–299

Remis JP, Costerton JW, Auer M (2010) Biofilms: structures that may facilitate cell-cell interactions. ISME J 4(9):1085–1087. doi:10.1038 (isme)

Romero R, Schaudinn C, Kusanovic JP, Gorur A, Gotsch F, Webster P, Nhan-Chang C-L, Erez O, Kim CJ, Espinoza J, Goncalves LF, Vaisbuch E, Mazaki-Tovi S, Hassan S, Costerton JW

(2008) Detection of a microbial biofilm in intraamniotic infection. Am J Obstet Gynecol 198:135.e1–135.e5

Sauer K, Camp AK, Ehrlich GD, Costerton JW, Davies DG (2002) *Pseudomonas aeruginosa* displays multiple phenotypes during development as a biofilm. J Bacteriol 189:1140–1154

Sauer K, Cullen MC, Rickard AH, Zeef LAH, Davies DG, Gilbert P (2004) Characterization of nutrient-induced dispersion in Pseudomonas aeruginosa biofilm. J Bacteriol 186:7312–7326

Singh P (2011) Biofilm driven evolution of a fitness trade off yields culture resistant bacteria. Eurobiofilms 2011, Copenhagen, Denmark, July 7 2011

Trampuz A, Piper KE, Jacobson MJ, Hanssen AD, Unni KK, Osmon DR, Mandrekar JN, Cockerill FR, Stekelberg JM, Greenleaf JF, Patel R (2007) Sonication of removed hip and knee prostheses for diagnosis of infection. N Engl J Med 357:654–663

Whitchurch CB, Tolker-Nielsen T, Ragas PC, Mattick JS (2002) Extracellular DNA required for bacterial biofilm formation. Science 295:1487

Wolcott RD, Rhoads DD, Bennett ME, Wolcott BM, Gogokhia L, Costerton JW, Dowd SE (2010) Chronic wounds and the medical biofilm paradigm. J Wound Care 19:45–54

# Culture-Negative Infections in Orthopedic Surgery

**G.D. Ehrlich, Patrick DeMeo, Michael Palmer, T.J. Sauber, Dan Altman, Greg Altman, Nick Sotereanos, Stephen Conti, Mark Baratz, Gerhard Maale, Fen Z. Hu, J. Christopher Post, Laura Nistico, Rachael Kreft, Luanne Hall-Stoodley, J.W. Costerton, and Paul Stoodley**

**Abstract** Laboratory cultures are the main scientific input into the decision-making process that determines the course of treatment for suspected orthopedic infections, just as they constitute the mainstay of the diagnosis of infections in other medical specialties. This situation is archaic because culture techniques were virtually abandoned in Environmental Microbiology (Hugenholtz et al. 1998) many years ago, following the conclusion that $<1$ % of the bacteria in any natural ecosystem can be recovered by standard cultural methods. Medical Microbiology has clung to

---

Editor's Note: The six chapters immediately following this insertion were first presented, in lecture form, at a conference entitled "Beyond cultures: the future of diagnostics in orthopedic infections," in Pittsburgh, PA between May 13 and 15, 2011. The contents of the lectures were converted to chapter form, with the capable editorial assistance of Dr. Dawn Marcus (MD), and each was approved and modified (as necessary) by the authors.

G.D. Ehrlich (✉) • F.Z. Hu • J.C. Post • L. Nistico • R. Kreft
Center for Genomic Sciences, Allegheny-Singer Research Institute, 320 E. North Avenue, Pittsburgh, PA, USA
e-mail: gehrlich@wpahs.org

P. DeMeo • M. Palmer • T.J. Sauber • D. Altman • G. Altman • N. Sotereanos • S. Conti • M. Baratz
Department of Orthopedic Surgery, Allegheny General Hospital, Pittsburgh, PA, USA

G. Maale
Dallas Ft. Worth Sarcoma Group, Dallas, TX, USA

L. Hall-Stoodley • P. Stoodley
University of Southampton, Southampton, UK

J.W. Costerton
Department of Orthopedic Surgery, Allegheny General Hospital, Pittsburgh, PA, USA

Center for Genomic Sciences, Allegheny-Singer Research Institute, 320 E. North Avenue, Pittsburgh, PA, USA

G.D. Ehrlich et al. (eds.), *Culture Negative Orthopedic Biofilm Infections*, Springer Series on Biofilms 7, DOI 10.1007/978-3-642-29554-6_2,

culture techniques because they detect the bacteria that cause acute infections, with reasonable sensitivity and accuracy, but the time has come to examine both their sensitivity and their accuracy for the detection and identification of bacteria in chronic biofilm infections (Costerton et al. 1999).

# 1 Culture Methods for Bacterial Detection and Identification

Cultures represent an arcane, but heretofore very useful, technology that remains largely unchanged since they were adopted in Berlin in the mid-1800s (Koch 1884). In this technique, body fluids or tissue specimens are taken from the patient and spread on the surface of an agar medium or grown in a nutrient broth designed to encourage the replication of bacterial cells, until millions of cells of that species either form a macroscopic "colony" on the agar plate or grow planktonically in the broth. The shape and color of the colony help to identify the species of bacteria that have formed the colony, the number of colonies is roughly proportional to the number of cells of that species in the original specimen, and biochemical tests are used to confirm the species identity of the "isolate." The colony can then be "picked" and grown in the presence of antibiotics, at various concentrations, to determine the antibiotic sensitivities of the strain concerned.

This FDA-approved method for the diagnosis of bacterial infections is predicated on the assumption that all of the bacteria of interest to the clinician will grow on the medium that is used, and on the assumption that every bacterial cell will give rise to a separate and distinct colony on the agar surface. In acute bacterial diseases (e.g., "strep throat") these assumptions are reasonable, and even 1/100 dilutions of the specimen will produce hundreds of characteristic colonies that fill the plate and provide both an unequivocal diagnosis and a basis for determining antibiotic sensitivity. The contemporary problem in modern medicine arises from the fact that as many as 80 % of all infections treated by physicians in the developed world are not caused by planktonic bacteria, but are caused by bacteria growing in slime-enclosed biofilms (Costerton et al. 1999). Of special interest to Orthopedic Surgery is the further revelation that virtually all device-related bacterial infections are caused by the biofilm form of bacterial growth (Khoury et al. 1992). The most intuitive of the problems in detecting biofilm bacteria by culture methods derives from the fact that the cells within biofilm communities are bound together by a viscous polysaccharide matrix so that they occur in coherent multicellular aggregates in the specimen (Fig. 1). Obviously hundreds, or even thousands, of bacterial cells bound together in an aggregate will give rise to only a single colony on the agar surface. Robin Patel's group at the Mayo Clinic addressed this problem of bacterial aggregation in orthopedic specimens (Trampuz et al. 2007) and showed that simple sonication breaks up biofilm aggregates, produces some single planktonic cells, and increases the proportion of putatively infected orthopedic prostheses that yield positive cultures.

The aggregation problem is significant, but is dwarfed by the problem posed by the failure of many bacterial cells to grow on the surfaces of the agar media used in

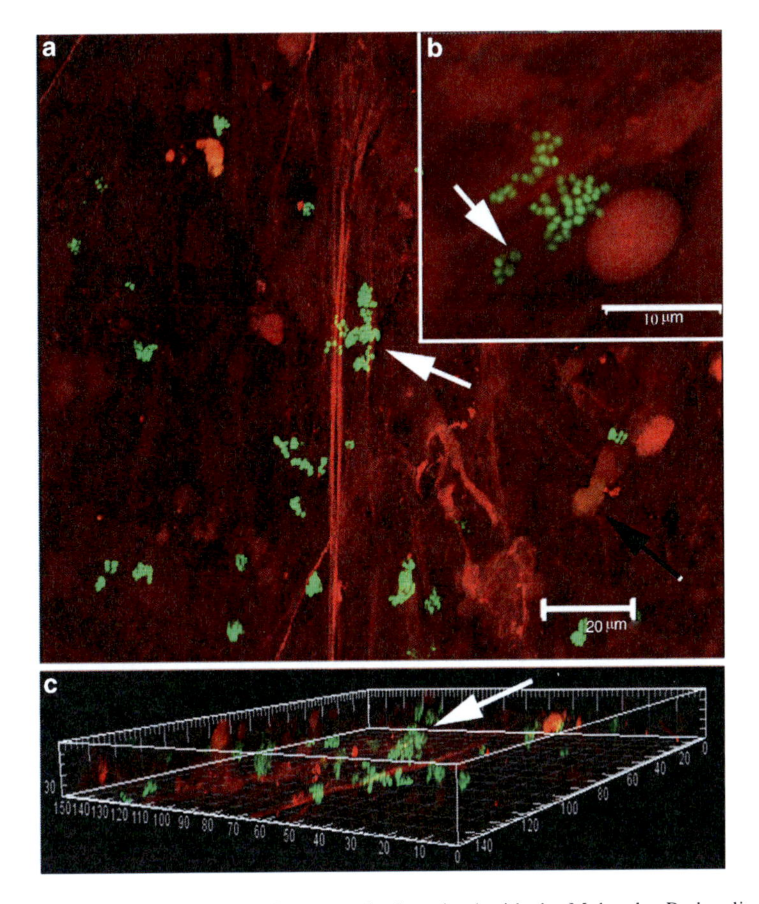


**Fig. 1** Material from an infected elbow prosthesis stained with the Molecular Probes live/dead viability kit and examined by confocal scanning laser microscopy (CSLM). Living cells (*green*) of *Staphylococcus aureus* (identified by PCR techniques) are sometimes present as single cells, but they usually form the clusters characteristic of this genus. The clusters (*white arrows*) of bacterial cells are seen in the *x–y* projection (A), the high magnification inset (B), and the three-dimensional orthogonal projection (C), and human material (*black arrow*) is seen in the background because it reacts with the propidium iodide in the live/dead kit

routine culture protocols. These media were designed to facilitate the growth of the small number of species that are called "professional pathogens," and that cause the preponderance of acute and epidemic infections in human beings. Most bacterial species fail to grow on the laboratory media used in routine culture protocols, and this phenomenon has led to the virtual abandonment of culture methods in Environmental Microbiology (Hugenholtz et al. 1998). Certainly most anaerobic bacteria cannot produce colonies in routine cultures, and bacteria with fastidious nutrient requirements (e.g., *Propionibacterium acnes*) remain undetected unless special isolation procedures are followed.

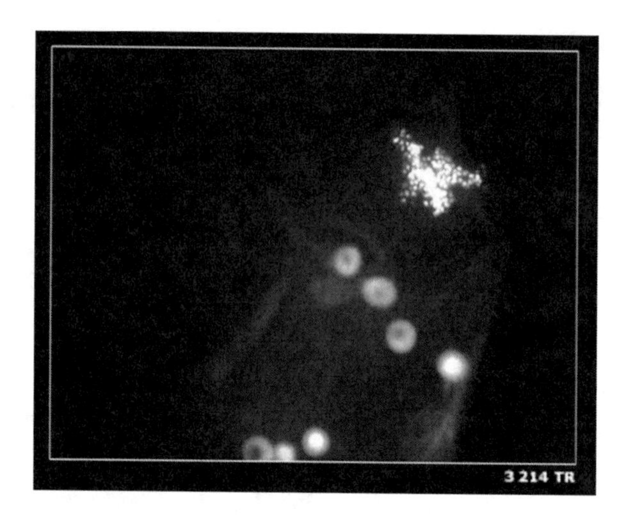


**Fig. 2** Fluorescence micrograph of an epithelial cell recovered from the vagina of a healthy human volunteer and reacted with a bacterial 16S FISH probe specifically designed to hybridize only with cells of *Staphylococcus aureus*. This unequivocal evidence of the presence of *S. aureus* biofilms on the vaginal epithelium of 100 % of 300 normal volunteers contrasts sharply with the finding that culture techniques only detected this organism in 10.8 % of the 3,000 volunteers examined in the original study. We concluded that 89.2 % of normal human volunteers, who are heavily colonized by *S. aureus* biofilms, do not yield positive cultures, even when cultures are taken and processed under optimal conditions (Veeh et al. 2003)

But the main cause to failure to grow on routine culture media, and of the consequent failure of culture methods to detect bacterial pathogens, is the fact that bacteria growing in biofilms simply do not produce colonies when they are transferred to the surfaces of agar plates. This phenomenon was first noted when we surveyed 3,000 volunteers for the presence of cells of *Staphylococcus aureus* in their vaginal flora (Veeh et al. 2003), and determined that 10.8 % yielded positive cultures, when swabs were transferred to the laboratory at body temperature and cultured immediately. We then examined vaginal scrapings from a subset of 300 of these volunteers, by modern molecular methods involving the detection of *S. aureus* cells by species-specific fluorescence *in situ* hybridization (FISH) probes (Fig. 2), and found that 100 % were heavily colonized by this organism. FISH probes are unequivocal because they rely on hybridization with prokaryotic (bacterial) 16 S rRNA, which does not occur in human tissues, so we must conclude that all women have this species in their vaginal flora but that only 10.8 % yield positive cultures. Subsequent longitudinal analyses by culture from a subset of the volunteers showed that positive cultures were entirely sporadic and random, leading to the suggestion (Veeh et al. 2003) that cultures were only positive when the coherent *S. aureus* biofilms (Fig. 2) shed planktonic cells that would produce colonies on agar plates. This suggestion was further reinforced when we showed that vaginal epithelial cells bearing large biofilm colonies of *S. aureus* failed to produce colonies when plated on agar media (unpublished data).

## 2 The Failure of Culture Methods in the Detection of Chronic Biofilm Infections

The simplest and most unequivocal instance of failure to culture, even when bacterial biofilms are present, involves a series of orthopedic device infections. When Sulzer Medical omitted a nitric acid cleaning step in the manufacture of their acetabular cup, several hundred complications known collectively as "aseptic loosenings" occurred. In this condition the prosthesis became loose, and there were multiple symptoms of bacterial infection, but aspirates and intraoperative specimens were uniformly culture negative: hence the name of "aseptic loosening." We (Maale and Costerton) examined eight consecutive explanted culture-negative acetabular cups, and the associated hardware within these hip prostheses, using modern molecular methods for the detection of bacteria. Very large numbers of bacteria were seen when we stained the "ingrowth" tissues at the edges of the acetabular cup with acridine orange (Fig. 3), and these organisms were aggregated in a pattern that proved that they had formed extensive biofilms in these tissues.

Further studies of these culture-negative explanted acetabular cups, using scanning electron microscopy and species-specific FISH probes, showed that the plastic cups were heavily colonized with spherical bacteria cells of *Staphylococcus epidermidis* (Fig. 4) and that the cells reacted with the species-specific 16S FISH probe for *S. epidermidis* (Fig. 4, inset).

The elbow prosthesis shown in Fig. 1 was removed from a patient who endured seven surgical procedures over the course of 5 years, beginning with the placement

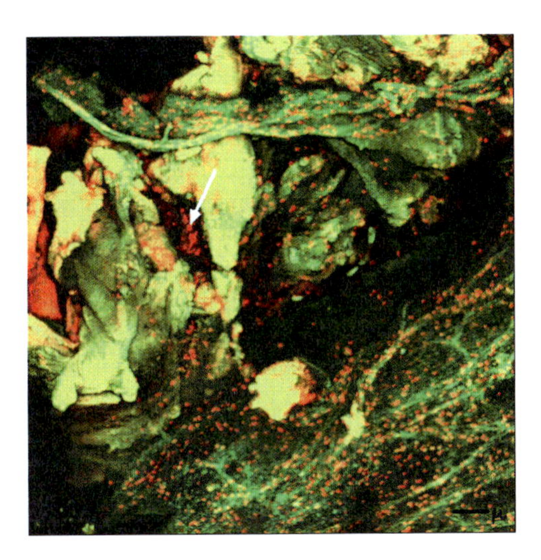

**Fig. 3** Confocal light micrograph of an acridine orange stained preparation of "ingrowth" tissue scraped from the surface of a culture-negative Sulzer acetabular cup. Hundreds of orange-stained bacteria are seen to have colonized some elements of the tissue, and a well-developed biofilm aggregate (*arrow*) is seen to fill one of the spaces between tissue components. *Bar* indicates 10 μm

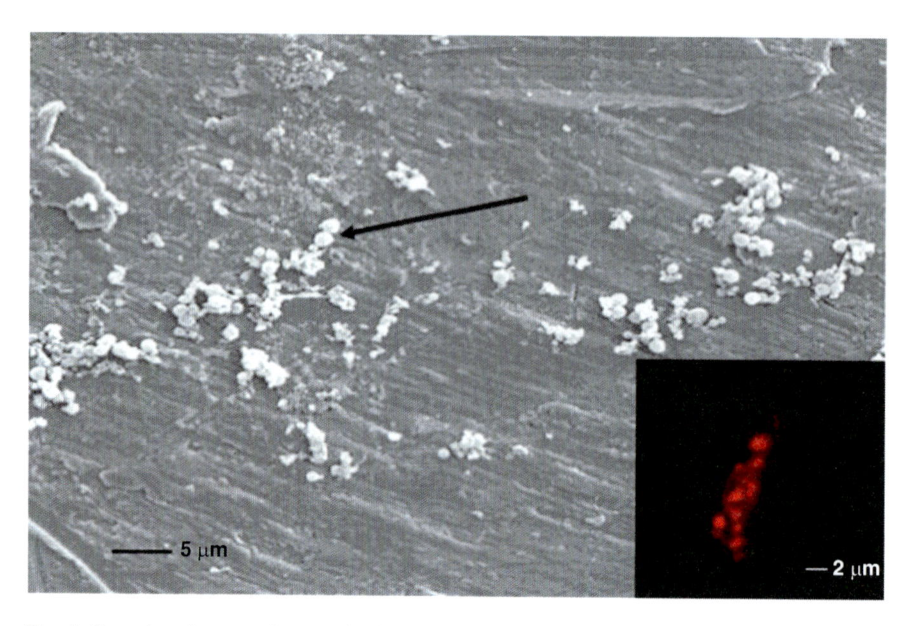


**Fig. 4** Scanning electron micrograph of a Sulzer acetabular cup, removed from a case of "aseptic loosening," showing the presence of spherical bacterial cells (*arrow*) in slime-enclosed clusters on the surface of the plastic. The inset shows that these spherical cells react with the species-specific 16S FISH probe for *S. epidermidis*

of a prosthesis and culminating with the removal of this device and the subsequent removal of an associated mass of methyl methacrylate. During this entire period, cultures were consistently negative while the patient's symptoms and the radiolucency of bone in X-rays (Stoodley et al. 2008) clearly indicated that a chronic bacterial infection had caused the failure of this trauma repair. Cultures were positive for *S. aureus* when intraoperative material was sent to the laboratory, from the final surgery, which left the patient with a "flail arm." This case represents a landmark in the diagnosis of chronic bacterial infections in Orthopedic Surgery because aspirates and tissue samples yielded positive results for the presence of *S. aureus* when examined by the reverse transcriptase polymerase chain reaction (RT-PCR), while cultures were consistently negative. We have continued to study individual cases in which cultures of aspirates and of intraoperative materials have been negative, while the attending surgeon suspected the presence of an infection, based on the patient's symptoms and on radiography. Figure 5 shows a bacterial biofilm on the plastic component of a prosthetic ankle, from a patient whose aspirates and intraoperative specimens were culture-negative, but whose prosthesis was clearly infected by biofilm-forming bacteria. This image shows the presence of living bacterial cells, while parallel analysis by PCR-ESI-TOF-MS showed the presence of *S. aureus*, and FISH probe analysis of the same sample showed the presence of large numbers of cells of this organism.

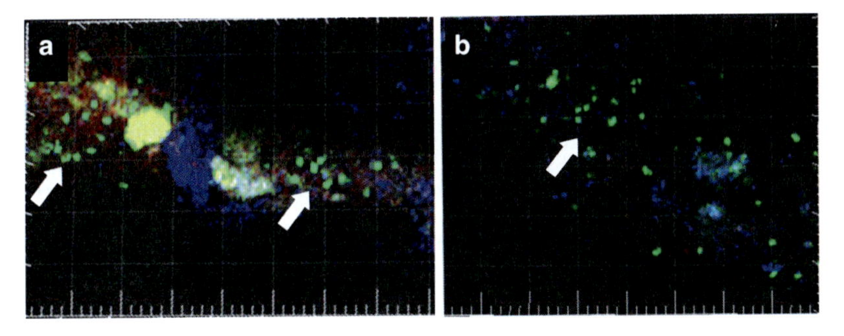


**Fig. 5** Confocal micrograph of a bacterial biofilm that had formed on the plastic component of a prosthetic ankle joint, which never yielded positive cultures, from aspirates or from intraoperative materials. The preparation has been reacted with the Molecular Probes live/dead kit so that living bacteria are *yellow/green* and dead bacteria are *red*, and these bacterial cells are seen to comprise an extensive matrix-enclosed microbial community in which most of the bacteria are alive. The *bar* indicates 10 μm

## 3 The Solution to the Problem of Negative Cultures

When microbial ecologists were faced with the problem of the lack of sensitivity and accuracy of classic culture techniques, in bacterial population analyses in natural ecosystems, they turned to DNA-based molecular techniques (Hugenholtz et al. 1998). Two streams of techniques soon developed, in that broad low-resolution techniques (e.g., DGGE and D-HPLC), were used to determine how many species were present in mixed populations, while focused high-resolution methods (e.g., DNA sequencing) were used to identify individual species. The molecular bases of these broad and focused techniques are discussed, in detail, in our recent review in FEMS Immunology and Medical Microbiology (Costerton et al. 2010). The DGGE technique was used to determine how many bacterial species were present in chronic wounds (James et al. 2008), identifying as many as 22 bacterial species in wounds (diabetic foot ulcers) that yielded positive cultures only for one or two commonly isolated pathogens (e.g., *S. aureus*). None of these molecular methodologies could have met the requirements of clinical diagnostic facilities, between 1980 and 2007, because none could provide rapid data and because the vital component of antibiotic resistance patterns was still lacking.

While they could not yet provide a routine technical platform for rapid and accurate diagnostics, DNA-based molecular techniques were used to establish the bacterial etiology of otitis media with effusion (OME), and several other chronic biofilm infections. Ehrlich and Post proved that large amounts of bacterial DNA were present in OME, although most cultures were negative (Post et al. 1995), and they even showed that the infected tissues contained short-lived bacterial messenger RNA (mRNA), to establish that these bacteria were alive and metabolically active (Rayner et al. 1998). Ultimately, these infections were established as classic biofilm infections when they displayed matrix-enclosed bacterial colonies upon imaging, in both animal models of infections (Ehrlich et al. 2002) and human middle-ear infections (Hall-Stoodley et al. 2006). Cultures in both of these studies were usually negative Dowd et al. (2008) have used a combination of PCR methods with pyro-sequencing to show that many bacteria and fungi are present simultaneously in chronic wounds, but that cultures only detect a small fraction of these pathogens. This information has produced dramatic improvements in treatment (Wolcott and Ehrlich 2008) because antibiotic therapy can now be directed at the control of all of the pathogens (e.g., *Candida albicans*), and all can be suppressed or killed. These studies suggest that the suppression of one pathogen amongst many may lead to the resurgence of the organisms that have not been detected or treated, and to the prolongation of infections that are already chronic and refractory.

All of these studies of device-related and other chronic bacterial infections have produced a burgeoning mass of evidence that culture methods are both inaccurate and insensitive in the diagnosis of bacterial infections. This realization has crystallized, in fields as dissimilar as ENT and Orthopedics, but cultures have persisted as the gold standard because they can (ideally) provide an answer in 24 h and an antibiogram in 48 h. Some PCR methods (Cloud et al. 2000), and some methods based on antibodies (Brady et al. 2006), provide very rapid diagnosis, but we only “find what we are looking for” and we do not get a global picture of all of the organisms that are present. So clinical medicine is poised and waiting for a method for the accurate and sensitive detection and identification of bacterial pathogens, and an equally accurate means of determining their sensitivity to antibiotics.

## 4 The IBIS PLEX-ID

The bioterrorism defense community has an urgent need for the rapid and accurate detection and identification of bacterial pathogens, and they have fostered the development of the Ibis universal biosensor. This technology is based on the “weighing,” by mass spectroscopy (Ecker et al. 2008), of PCR-amplified bacterial DNA in samples, and the use of a complex algorithm to match the base ratios in these amplicons with those of many hundreds of bacterial species whose base ratios are stored in an interactive database (Fig. 6).

We have begun to compare the diagnostic capabilities of the IBIS technology with those of routine cultures, and, while we cannot yet predicate our treatment based on the Ibis system because of FDA restrictions, the advances in sensitivity and accuracy are patently obvious. Table 1 shows the data from four cases that were analyzed before we started our very extensive blinded prospective clinical trials of putative infections of total joints and of infected nonunions.

In each case in which culture methods had produced a diagnosis (MRSA or MRSE), the Ibis technique confirmed that diagnosis by detecting *S. aureus* or *S. epidermidis*, and the *Mec A* methicillin resistance gene cassette. In all of these positive culture cases, the Ibis detected additional organisms, and, in cases 010609 and 122308, these data would have changed the strategy for antibiotic therapy. In the

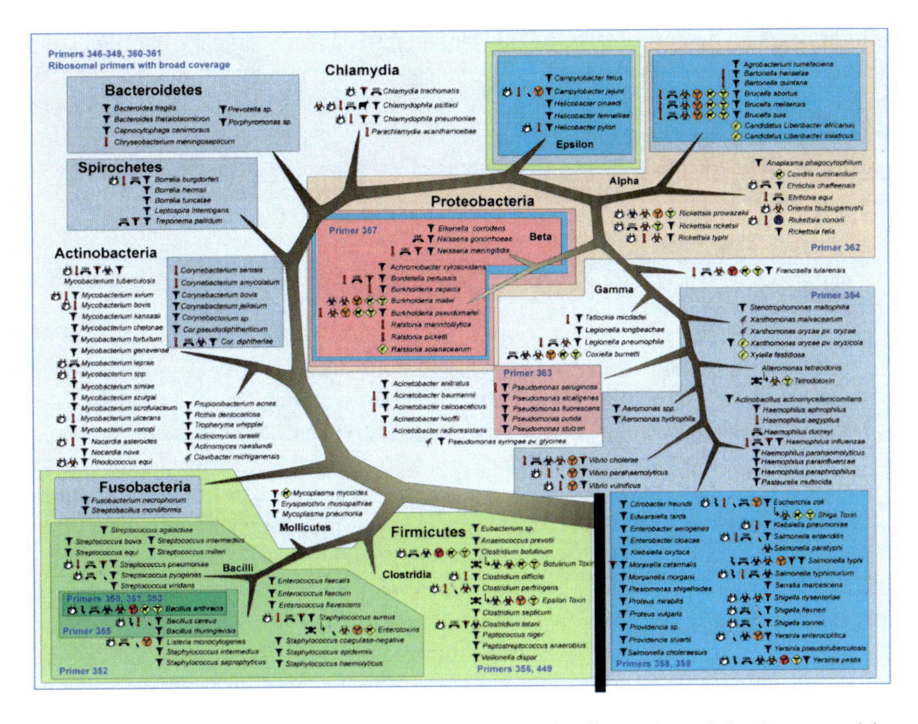


**Fig. 6** The IBIS database contains the base ratios of thousands of known bacterial pathogens, and the base ratios of the bacteria in samples can be determined and matched to those in the database, to determine the presence of any of these organisms. If an organism is not present in the database, it will be detected but not identified, and the relative prevalence of all organisms is established by the number of genomes present in the sample. The Ibis system also detects the genes responsible for antibiotic resistance, so a molecular antibiogram is provided in the 6-h time frame necessary for this analysis

**Table 1** Orthopedic cases in which the IBIS technique was compared with routine cultures

| Ortho case | Culture/Gram stain | Amplicon number and confidence of match | Prevalence = genomes/well | Identification of bacteria + antibiotic resistance |
|---|---|---|---|---|
| 120308 | MRSA | 1 = 100 % | 3,889 | *S. aureus* |
| | | 2 = 92.4 % | 452 | *S. epidermidis* |
| | | 3 = 100 % | 8,184 | Methicillin Res. |
| 121908 | Culture neg. | 1 = 90.8 % | 10,739 | *S. warneri* |
| | Few Gram + cocci | 2 = 100 % | 11,429 | *S. capitis* |
| | | 3 = 88.2 % | 1,460 | *P. acnes* |
| | | 4 = 100 % | 1,777 | Methicillin Res. |
| 010609 | MRSA | 1 = 87.9 % | 267 | *S. aureus* |
| | | 2 = 96.7 % | 124 | *S. epidermidis* |
| | | 3 = 94.8 % | 641 | *E. faecalis* |
| | | 4 = 99.7 % | 7,474 | *B. cereus* |
| | | 5 = 100 % | 714 | Methicillin Res |
| 122308 | MRSE | 1 = 99.3 % | 6,315 | *S. epidermidis* |
| | | 2 = 97.6 % | 2,058 | *S. capitis* |
| | | 3 = 99.6 % | 900 | *B. cereus* |
| | | 4 = 100 % | 20,236 | Methicillin Res. |

culture-negative case, very large numbers of methicillin-resistant coagulase-negative Staphylococci were found by the Ibis technology, and a chronic biofilm infection was clearly present. In this case a Gram-positive pathogen (*P. acne*) was also present and would have triggered appropriate antibiotic therapy if the IBIS system was approved for bacterial diagnosis.

## 5 Summary

Culture methods are no longer used for the detection and identification of bacteria, in many fields of Microbiology (e.g., Microbial Ecology) that can accommodate the leisurely pace of pyro-sequencing and other DNA-based molecular methods. Where these methods have been used to detect and identify bacteria in human infections, they have proven to be more accurate and more sensitive than culture methods, but their slow pace and high cost have prevented their adoption for routine diagnosis. Some very rapid diagnostic methods, based on PCR amplification or on reaction with specific antibodies, have gained some acceptance, but these highly focused methods only look for specific organisms and cannot detect all of the bacteria present in a sample. We have examined a new mass-spec-based technology for the detection and identification of bacteria that is based on the base ratios of segments of several critical bacterial genes, and that is very rapid (<6 h) because it does not involve sequencing of these bases. This Ibis technology detects and identifies bacteria with much more sensitivity than cultures, and it solves the dual problems posed by biofilms, in that cells in clusters are detected individually, and in that cells that fail to grow on culture media are detected quantitatively. The Ibis technology also detects the major bacterial genes that control antibiotic resistance, so bacteria can be detected and identified in 6 h, and their antibiotic resistance profiles are also known in this same short timeframe. We suggest that a systematic comparison of the Ibis technology with culture methods should be undertaken, with full training of clinicians in the interpretation of molecular data, with the intent of replacing cultures with molecular techniques in the immediate future.

## References

Brady RA, Leid JG, Camper AK, Costerton JW, Shirtliff ME (2006) Identification of *Staphylococcus aureus* proteins recognized by the antibody-mediated immune response to a biofilm infection. Infect Immun 74:3415–3426

Cloud JL, Carroll KC, Pixton P, Erali M, Hillyard DR (2000) Detection of Legionella species in respiratory specimens using PCR with sequencing confirmation. J Clin Microbiol 38:1709–1712

Costerton JW, Stewart PS, Greenberg EP (1999) Bacterial biofilms: a common cause of persistent infections. Science 284:1318–1322

Costerton JW, Post C, Ehrlich GD, Hu FZ, Kreft R, Nistico L, Kathju S, Stoodley P, Hall-Stoodley L, Maale G, James G, Shirtliff M, Sotereanos N, DeMeo P (2010) New rapid and accurate

methods for the detection of orthopaedic infections. FEMS Immunol Med Microbiol 61:133–140
Dowd SE, Wolcott RD, Sun Y, McKeehan T, Smith E, Rhoads D (2008) Polymicrobial nature of chronic diabetic foot ulcers using bacterial Tag encoded amplicon pyro-sequencing (bTEFAD). PLoS One 3:e3326
Ecker DJ, Sampath R, Massire C, Blyn LB, Hall TA, Eshoo MW, Hofstadler SA (2008) Ibis T5000: a universal biosensor approach for microbiology. Nat Rev Microbiol 6:553–558
Ehrlich GD, Veeh R, Wang X, Costerton JW, Hayes JD, Hu FZ, Daigle BJ, Ehrlich MD, Post JC (2002) Mucosal biofilm formation on middle-ear mucosa in the chinchilla model of otitis media. JAMA 287:1710–1715
Hall-Stoodley L, Hu FZ, Gieseke A, Nistico L, Nguyen D, Hayes J, Forbes M, Greenberg DP, Dice B, Burrows A, Wackym PA, Stoodley P, Post JC, Ehrlich GD, Kerschner JE (2006) Direct detection of bacterial biofilms on the middle ear mucosa of children with otitis media. J Am Med Assoc 296:202–211
Hugenholtz P, Goebel BM, Pace NR (1998) Impact of culture-independent studies on the emerging phylogenetic view of bacterial diversity. J Bacteriol 180:4765–4774
James GA, Swogger E, Wolcott R, deL Pulcini E, Secor P, Sestrich J, Costerton JW, Stewart PS (2008) Biofilms in chronic wounds. Wound Repair Regen 16:37–44
Khoury AE, Lam K, Ellis B, Costerton JW (1992) Prevention and control of bacterial infections associated with medical devices. ASAIO Trans 38(3):M174–M178
Koch R (1884) Die aetiologie der tuberkulose, mittheilungen aus dem kaiserlichen. Gesundhdeitsamte 2:1–88
Post JC, Preston RA, Aul JJ, Larkins-Pettigrew M, Ridquist-White J, Anderson KW, Wadowsky RM, Reagan DR, Walker ES, Kingsley LA, Ehrlich GD (1995) Molecular analysis of bacterial pathogens in otitis media with effusion. J Am Med Assoc 273:1598–1604
Rayner MG, Zhang Y, Gorry MC, Chen Y, Post JC, Ehrlich GD (1998) Evidence of bacterial metabolic activity in culture-negative otitis media with effusion. J Am Med Assoc 279:296–299
Stoodley P, Nistico L, Johnson S, Lasko L-A, Baratz M, Gahlot V, Ehrlich GD, Kathju S (2008) Direct demonstration of viable *Staphylococcus aureus* biofilms in an infected total joint arthroplasty: a case study. J Bone Joint Surg Am 90:1751–1758
Trampuz A, Piper KE, Jacobson MJ, Hanssen AD, Unni KK, Osmon DR, Mandrekar JN, Cockerill FR, Stekelberg JM, Greenleaf JF, Patel R (2007) Sonication of removed hip and knee prostheses for diagnosis of infection. N Engl J Med 357:654–663
Veeh RH, Shirtliff ME, Petik JR, Flood JA, Davis CC, Seymour JL, Hansmann MA, Kerr KM, Pasmore ME, Costerton JW (2003) Detection of Staphylococcus aureus biofilm on tampons and menses components. J Infect Dis 188:519–530
Wolcott RD, Ehrlich GD (2008) Biofilms and chronic infections. J Am Med Assoc 299:2682–2684

# Improved Diagnosis of Biofilm Infections Using Various Molecular Methods

**Trine Rolighed Thomsen, Yijuan Xu, Jan Lorenzen, Per Halkjær Nielsen, and Henrik Carl Schønheyder**

**Abstract** Traditional culture-dependent methods and a number of culture-independent molecular methods including 16S rRNA gene polymerase chain reaction, construction of clone libraries, sequencing, phylogeny, fingerprinting, fluorescence *in situ* hybridization and quantitative PCR were used to describe the microbial composition of two types of biofilm-related infections, namely chronic venous leg ulcers and prosthetic joint infections. Multiple tissue biopsies were taken from each chronic wound, and different specimen types (joint fluid, tissue biopsy, bone biopsy and prosthesis scraping or sonication) were collected from prosthetic joint patients. The obtained results indicate that in these two types of infections the bacterial composition and yield may vary depending on the position and type of samples used for analysis. It emphasizes the need for multiple samplings in order to achieve better diagnosis and treatment of these biofilm-related infections. The most complete picture of microbial composition of biofilms is probably accomplished when several culture and culture-independent methods are used in parallel to characterize the pathogens.

T.R. Thomsen (✉) • Y. Xu
Department of Biotechnology, Chemistry, and Environmental Engineering, Aalborg University, Aalborg, Denmark

Life Science Division, The Danish Technological Institute, Aarhus, Denmark
e-mail: trt@bio.aau.dk

J. Lorenzen
Life Science Division, The Danish Technological Institute, Aarhus, Denmark

P.H. Nielsen
Department of Biotechnology, Chemistry, and Environmental Engineering, Aalborg University, Aalborg, Denmark

H.C. Schønheyder
Department of Clinical Microbiology, Aalborg Hospital, Aarhus University Hospital, Aalborg, Denmark

G.D. Ehrlich et al. (eds.), *Culture Negative Orthopedic Biofilm Infections*,
Springer Series on Biofilms 7, DOI 10.1007/978-3-642-29554-6_3,

# 1 Introduction

Bacteriological cultures often underestimate the pathogens present in chronic infections. This is often due to a combination of inadequate growth conditions and the presence of slow, fastidious, anaerobic, or unculturable bacteria growing in biofilms. Molecular techniques have been proven to be effective in demonstrating bacterial diversity in a broad range of environmental samples, including activated sludge, oil systems, district heating systems, drinking water, and medical samples (e.g., Amann et al. 1990; Drancourt et al. 2000; Kjellerup et al. 2005; Burmølle et al. 2010). In clinical medicine, molecular techniques are often able to identify less common pathogens that may not grow readily on laboratory culture media.

While molecular techniques offer clear advantages for characterizing chronic biofilm infections, it is important to understand how to best apply these techniques and why using a variety of techniques in combination is often more effective than a single technique. As with any diagnostic, it is important to understand the potential shortcomings of molecular techniques and be aware of improvements that can reduce some of these limitations. Reliable results depend on adequate sample collection and interpretation depends on awareness that pathogens are frequently distributed heterogeneously through biofilms. This chapter will focus on the application of molecular techniques to chronic wounds and orthopedic infections.

# 2 Overview of Molecular Techniques for Chronic Infection Evaluation

A wide range of molecular techniques are currently being evaluated for the diagnosis of chronic infections (Box 1). Each of these techniques has been successfully applied to chronic wound evaluation (Thomsen et al. 2010). Importantly, data show that results often vary among different techniques, supporting inclusion of several molecular techniques for clinical evaluation rather than relying on a single method.

## *2.1 Polymerase Chain Reaction*

Polymerase chain reaction (PCR) diagnosis of chronic infections is described in detail in Chapter 4 (Kennedy). PCR permits DNA-identification of pathogens by amplifying bacterial DNA that matches pre-selected pathogen primer sequences.

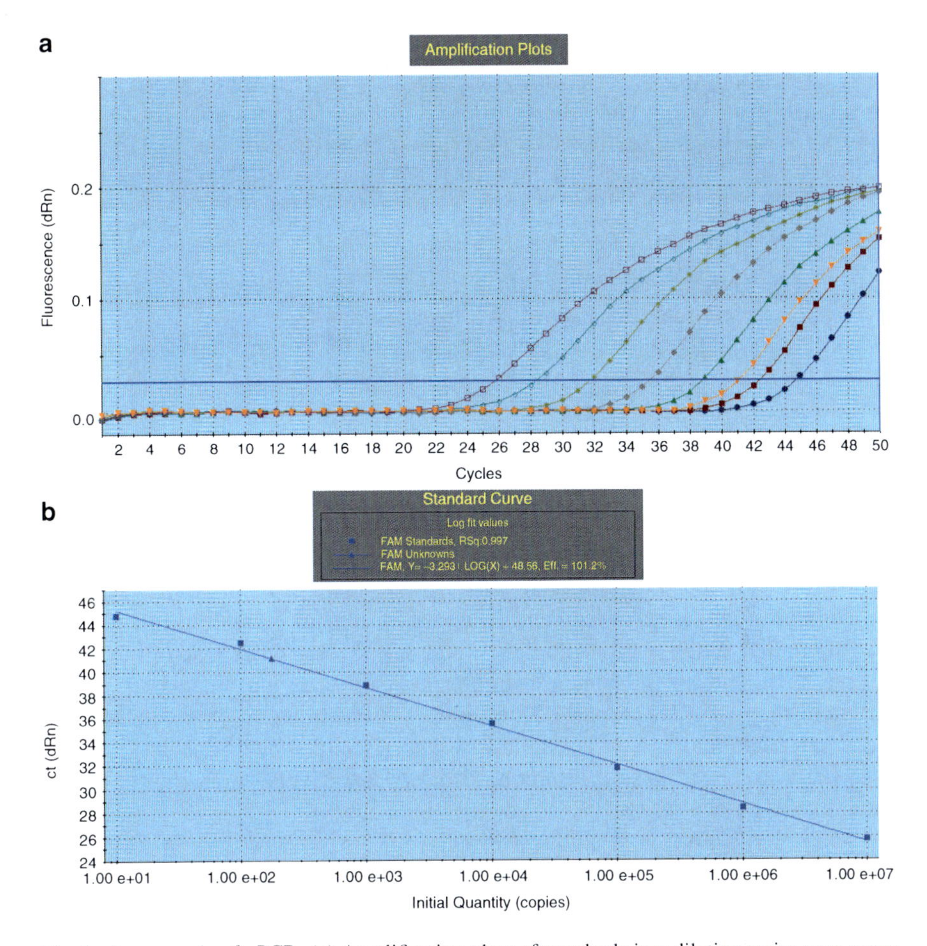


**Fig. 1** An example of qPCR. (**a**) Amplification plots of standards in a dilution series over seven orders of magnitude and one unknown sample (*orange*). The *horizontal blue line* represents the threshold value. The PCR cycle at which a sample's fluorescence reaches the threshold value defines the threshold cycle (Ct) of the sample. The Ct is inversely proportional to the log of the initial copy number of target gene. (**b**) Standard curve generated with data from (**a**) with the unknown sample shown as a *triangle*

PCR with 16S ribosomal RNA (rRNA) gene sequencing has proven to be an important tool for identifying bacteria that may be difficult to grow in culture. 16S rRNA gene sequencing also helps identify potential novel pathogens that do not grow on standard culture media.

Quantification of bacterial density can be achieved through quantitative PCR (qPCR) testing, which measures fluorescence at each cycle as the amplification progresses. The exact amount of initial target gene in a sample can be calculated by

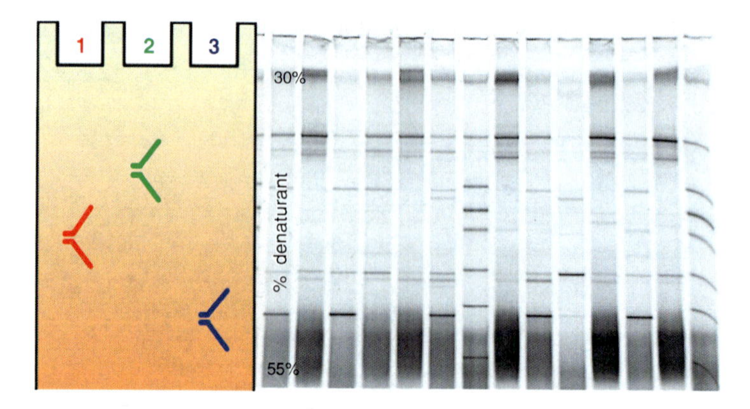


**Fig. 2** The principle and an example of DGGE. Each band can be cut out from the gel, reamplified, and sequenced. Gel is provided by Nordentoft and Bjerrum Friis-Holm; denaturant gradient was 30–55 % (unpublished results)

using a standard curve prepared from a dilution series of control templates of known concentration (Fig. 1).

Interpreting PCR, however, may be limited by various biases: primer specificity, differential amplification efficiency and incomplete databases for bioinformatics search, and sample contamination from dead bacteria or human host DNA. For example, 16S rRNA gene PCR testing has been shown to detect both viable and dead bacteria (Kobayashi et al. 2009). To circumvent this issue, some researchers have suggested detecting mRNA by reverse transcriptase PCR because mRNA can be used as a viability indicator (Hellyer et al. 1999; Klein and Juneja 1997; Zhao et al. 2006). However, mRNA was shown to be detectable for an extended period after cell death in some other studies (Birch et al. 2001; Sheridan et al. 1998; Sung et al. 2005). Depending on the bacterial species, mRNA target, inactivating treatment, and subsequent holding conditions, the stability of mRNA after bacterial death may vary (Sheridan et al. 1998, 1999; Sung et al. 2005; Yaron and Matthews 2002). Instead of targeting mRNA, selective isolation of DNA from intact bacteria (described below) may improve diagnosis from PCR techniques (Horz et al. 2008).

## 2.2 *Fluorescence* In Situ *Hybridization*

Fluorescence *in situ* hybridization (FISH) detects and localizes nucleic acid sequences on intracellular rRNA using oligonucleotide fluorescent probes. FISH images, therefore, provide information about both the presence and location of pathogens that may be widespread or confined to local areas. FISH may be conducted using DNA or peptide nucleic acid (PNA) technology. DNA's negatively charged sugar-phosphate backbone may produce undesirable electrostatic repulsion that impedes binding. PNA technology contains the same nucleotide bases as DNA, with the negatively charged backbone replaced with a noncharged peptide backbone to allow stronger and more rapid hybridization. Particularly in medical samples, the PNA technique often seems simpler and faster.

### 2.3 *Fingerprinting*

Denaturing gradient gel electrophoresis (DGGE) is the most common fingerprinting technique for identifying pathogens. Microbes are recognized, based on PCR-amplified nucleic acid fragments (Muyzer et al. 1993; Tzeneva et al. 2008). DNA sequences are separated using electrophoresis, and banding patterns can be used to provide a "fingerprint" for specific pathogens. Upon staining, each band on the gel theoretically represents a unique 16S rRNA gene amplicon that can be excised and extracted from the gel and sequenced for further identification (Fig. 2).

### 2.4 *Clone Library*

Genomic DNA from a mixture of microorganisms, including pathogens of interest, can be isolated and subdivided into clonable elements. The clone library method is frequently based on the 16S rRNA gene for species identification. The clone library approach can identify individual species in a polymicrobial infection, in contrast to 16S rRNA gene PCR followed by direct sequencing, which is unable to differentiate mixed 16S rRNA gene products derived from multiple species. The principle is that PCR-amplified segments are ligated into vectors and subsequently cloned into host cells to create a gene bank or clone library. The vectors are then extracted from each individual colony, and the 16S rRNA gene amplicons are sequenced. By performing Basic Local Alignment Search Tool (BLAST) searches in public databases (Altschul et al. 1997), the obtained sequences can be assigned to specific bacteria. An improved phylogenetic resolution can be obtained by performing a phylogenetic analysis on the basis of the sequences. iSentio (Norway) has recently developed RipSeq, a web-based application for the interpretation of chromatograms containing mixed 16S rRNA gene products derived from multiple species, which is highly relevant for routine diagnostic use (Kommedal et al. 2011). The clone library method can then be omitted in polymicrobial samples with relative low bacterial diversity.

## 3 Molecular Techniques Highlight Biofilm Diversity in Chronic Wounds

A broader diversity of bacteria is characteristically identified when molecular techniques are added to routine culture results. Davies and colleagues compared microflora in chronic venous leg ulcers using cultures and molecular techniques

**Table 1** Overview of molecular techniques

| Technique | Description |
|---|---|
| Cloning with sequencing and phylogenetic studies | Amplified primer-targeted gene sequences are cloned and sequenced. The sequences are used for constructing phylogenetic trees |
| Denaturing gradient gel electrophoresis fingerprinting (DGGE) | PCR amplified sequences are separated by providing them with a GC-rich primer tail and loading them on a polyacrylamide gel. Bands can be excised from the gel and sequenced |
| Quantitative PCR | Specific gene target is amplified by PCR and simultaneously quantified by applying a fluorescent reporter molecule |
| Fluorescence in situ hybridization | Identification and visualization of specific pathogens of interest by using fluorescent oligonucleotide probes to target rRNA |

*PCR* polymerase chain reaction

(PCR and fingerprinting; Davies et al. 2004). While cultures predominantly identified *Staphylococcus* and *Pseudomonas* species, molecular techniques revealed a broader range of pathogens. Over 40 % of the sequences represented microorganisms which had not been identified through routine culture.

## 3.1 Diversity Identification is Maximized by Evaluating Culture and Culture-Independent Techniques

The most complete picture of microbial makeup of biofilms is probably accomplished when a range of culture and culture-independent methods are used to characterize microorganisms. Both culture and molecular techniques were used to evaluate bacterial diversity in material from skin graft operations from 14 patients having chronic, nonhealing venous leg ulcers (Thomsen et al. 2010). Molecular techniques included 16S rRNA gene sequencing, cloning with sequencing and phylogenetic studies, DGGE, qPCR, and FISH (Table 1). Only three of the patients who provided samples had been using antibiotics during the 3 months before sample collection. Based on all applied methods it was concluded that each wound contained an average of 5.4 pathogens, with individual bacteria varying among wounds. Culture detection identified 12 different pathogens across analyzed wounds, predominantly *S. aureus*, *P. aeruginosa*, and *Enterococcus faecalis* (Table 2). Molecular methods, conversely, detected 33 pathogens, including a broad range of unexpected species and important anaerobes that were missed with cultures. All the wounds contained *S. aureus*, whereas *P. aeruginosa* was detected in 6 out of 14 wounds.

In some cases, pathogens identified through culture were not detected using molecular techniques. For example, while *S. aureus* and *P. aeruginosa* were predominantly identified across wounds, qPCR testing for these two organisms showed variability in pathogen abundance among wounds, with quantities sufficient to produce positive qPCR results for *S. aureus* in only 4 of the 14 wounds and *P. aeruginosa* in 3 wounds. These data reinforce the need to use information from both culture and culture-independent techniques to maximize pathogen identification.

**Table 2** Examples of wound pathogens detected by different techniques (based on Thomsen et al. 2010)

| Culture | Molecular techniques | |
|---|---|---|
| | Fingerprinting | Clone library |
| *Staphylococcus aureus* | *Staphylococcus aureus* | *Staphylococcus aureus* |
| *Pseudomonas aeruginosa* | *Pseudomonas aeruginosa* | *Pseudomonas aeruginosa* |
| *Klebsiella oxytoca* | *Finegoldia magna* | *Alcaligenes* species |
| *Enterococcus* species | *Anaerococcus vaginalis* | *Anaerococcus* species |
| | *Peptoniphilus asaccharolyticus* | *Stenotrophomonas* species |
| | *Peptoniphilus harei* | *Enterococcus faecalis* |
| | *Peptostreptococcus anaerobius* | |

### 3.2 Molecular Techniques Reinforce Sampling Requirements

Results with both culture and molecular techniques rely on appropriate wound sampling. Important differences in pathogen identification based on the location within the wound have been highlighted in several studies and the clinical relevance of this has been discussed (e.g., Fazli et al. 2009; Kirketerp-Møller et al. 2008; Wolcott et al. 2009). Wound cultures identified *S. aureus*, *E. coli*, and coagulase-negative *Staphylococcus lugdunensis* in a study of a chronic venous leg ulcer in a 59-year-old man by Andersen et al. (2007). Fingerprinting analysis showed important spatial variation in bacteria abundance, based on sample location. Fingerprinting additionally confirmed that *S. aureus* was the predominant wound pathogen, with *S. lugdunensis* not identified and considered to likely represent a wound surface contaminant.

The study comparing pathogen discovery using culture versus molecular techniques in 14 chronic venous ulcerleg wounds described above further highlighted the importance of appropriate wound sampling (Thomsen et al. 2010). Sampling in different parts of individual wounds identified substantial differences in pathogen abundance. Three wounds were divided into five regions, one in the center and four evenly spaced around the wound periphery. Samples from each region were analyzed separately. Using qPCR testing, *S. aureus* and *P. aeruginosa* were identified in each tested wound region; however, pathogen quantity was variable and inconsistent across regions within the same wound. For example, in one wound the quantity of *P. aeruginosa* varied among testing sites by more than three orders of magnitude (Table 3, wound 2). Similar heterogeneity in pathogen distribution was shown for *S. aureus*. FISH testing further confirmed that pathogens could be located in discrete regions, with bacteria not identified across broad regions of FISH images (data not shown).

## 4 Applying Molecular Techniques to Prosthetic Joint Infection

"Prosthesis: Reduction of Infection and Pain" (Danish acronym PRIS) is a cross-disciplinary consortium funded by the Danish Agency for Science, Technology, and Innovation to study infections and pain related to implanted joint prostheses from

**Table 3** The heterogeneous distribution of *P. aeruginosa* identified through qPCR in two wounds (based on Thomsen et al. 2010)

| Sample location | Wound 1 | Wound 2 |
|---|---|---|
| Center | 510 ± 18 % | 920 ± 9 % |
| Periphery 1 | No sample | 300 ± 13 % |
| Periphery 2 | 760 ± 7 % | 8,200 ± 8 % |
| Periphery 3 | 47± 9 % | 800 ± 10 % |
| Periphery 4 | 280 ± 3 % | 15 ± 5 % |

Numbers are gene copies/ng DNA ± standard error of the mean

2010 through 2013 (PRIS website). One of the objectives is to develop new molecular biology tools for identifying bacteria in prosthetic joint infections (PJI). Other objectives are optimization of culture methods and radionuclide imaging, extensive pain assessment, and development of new biomarkers. A pilot study has compared identification of bacteria in PJI using culture and culture-independent techniques.

## 4.1 *Methodology Considerations*

Tissue sampling through the consortium has been standardized (Table 4). Positive cultures were interpreted as recommended by Kamme and Lindberg: one or more identical pathogens were required to be present in at least three of five biopsies to indicate significant infection (Kamme and Lindberg 1981). This method has been confirmed to apply to infection diagnosis in arthroplasty, although relatively low sensitivity and negative predictive value support using other methods in addition to culture diagnosis (Mikkelsen et al. 2006).

Because the presence of human DNA may confound identification of important microbial genetic material, DNA extraction was conducted with the MolYsis™ selective isolation system. This system provides discriminating lysis of human cells and degradation of human DNA as well as naked bacterial DNA, thereby enhancing the sensitivity and specificity for detecting viable and hence more clinically relevant bacteria in the subsequent molecular analysis (Fig. 3). The MolYsis™ system has been shown to remove >90 % of human DNA from samples that contain both human and bacterial cells (Horz et al. 2010).

## 4.2 *Preliminary Results*

Preliminary data are available from a pilot study of patients with suspected PJI. In this study of 25 patients, 16S rRNA gene PCR was positive in five joint fluid samples, five synovial membrane biopsies, four bone biopsies, and nine prosthesis samples. Molecular techniques were more robust in identifying the presence of a bacterial etiology and a broader variety of pathogens in infections. Distribution of bacteria, however, was heterogeneous, with prosthesis samples providing the most abundant source of bacteria.

**Table 4** Standardized tissue sampling for PJIs within the framework of the Danish cross-disciplinary "Prosthesis: Reduction of Infection and Pain" consortium

| Procedure during which sample is procured | Samples for molecular analysis and culture |
|---|---|
| Joint fluid aspiration | Joint fluid |
| Arthroscopy | Synovial membrane biopsy |
| Revision arthroplasty (small components exchanged) | Joint fluid |
| | Synovial membrane biopsies |
| | Bone biopsies |
| | Prosthesis scraping swab |
| | Removed prosthesis components |
| Revision arthroplasty (large components removed) | Joint fluid |
| | Synovial membrane biopsies |
| | Bone biopsies |
| | Cement spacer or removed prosthesis |

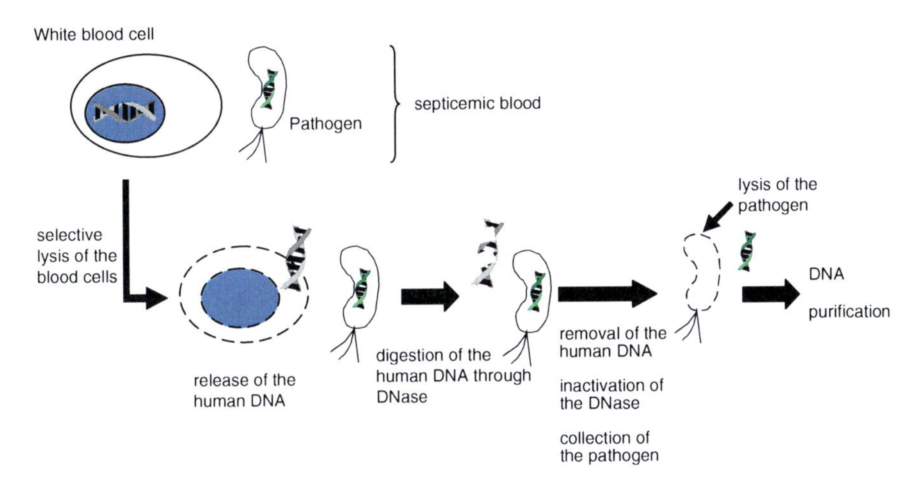


**Fig. 3** Optimized DNA extraction using MolYsis™ (reprinted with permission from Molzym, Germany)

Table 5 shows the results of qPCR testing in three representative patients across samples. These data underscore the need to include additional samples besides joint fluid analysis for an infection diagnosis.

## *4.3 Case Presentation*

Advantages and limitations from molecular testing can be highlighted by reviewing an individual case. An elderly man received a knee prosthesis 1 month prior to sample collection. Within an interval of 10 days, the patient underwent two revision

**Table 5** qPCR for *Propionibacterium*, numbers are average copies/μL ± the standard deviation

| Sample | qPCR results | | |
|---|---|---|---|
| | Patient 1 | Patient 2 | Patient 3 |
| Joint fluid | 0 | 0 | 0 |
| Bone | 43 ± 3 | 0 | 0 |
| Synovial membrane | 101 ± 95 | 39 ± 9 | 0 |
| Prosthesis scraping | 42 ± 14 | 0 | 83 ± 7 |

**Table 6** Organisms identified from a case of suspected PJI

| Culture | 16S rRNA gene analysis | | | |
|---|---|---|---|---|
| | Joint fluid | Synovial biopsy | Bone biopsy | Prosthesis sample |
| All five tissue samples were negative | *Streptococcus dysgalactiae* subspecies *equisimilis* | *Streptococcus dysgalactiae* subspecies *equisimilis* | *Streptococcus dysgalactiae* subspecies *equisimilis* | *Streptococcus dysgalactiae* subspecies *equisimilis* |
| | | | *Propionibacterium acnes* | *Propionibacterium acnes* |
| | | *Propionibacterium acnes* | | *Propionibacterium granulosum* |
| | | *Alcaligenes faecalis* | *Sphingomonas* species | Uncultured *Burkholderia* species |
| | | *Sphingomonas* species | *Streptococcus sanguinis* | |

Specimens were obtained during a second revision arthroplasty after 10 days of antibiotic treatment

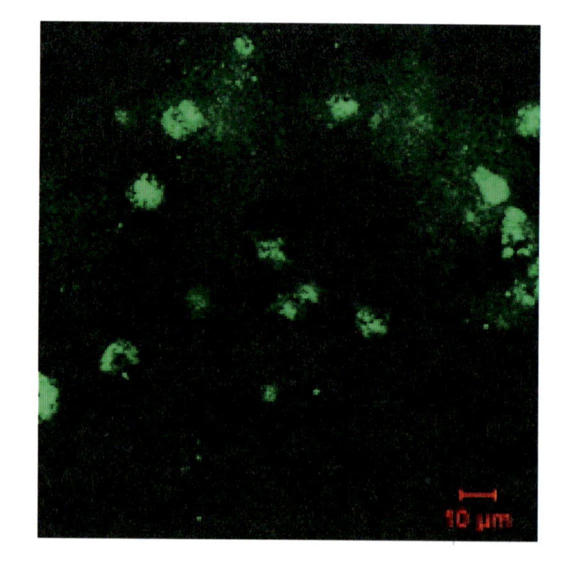


**Fig. 4** Fluorescent in situ hybridization (FISH) using a Cy5-labeled DNA-probe targeting all Gammaproteobacteria (Gam42a, Manz et al. 1992). Positive microcolonies were identified in sediment from a joint fluid sample. The scale bar represents 10 μm

arthroplasties with debridement and prosthesis retention; antibiotic treatment was administered between revisions. Cultures from tissue biopsies obtained during the first revision grew hemolytic streptococci serogroup G. Samples for both culture and molecular testing were available from the second revision. While all cultures were negative, 16S rRNA gene analysis identified a variety of organisms (Table 6). *Streptococcus dysgalactiae* subspecies *equisimilis* was found in all samples evaluated with molecular analysis. This taxon is compatible with the previous isolate of hemolytic streptococcus serogroup G. The other microorganisms detected by rRNA gene clone library were heterogeneous among samples, reinforcing recommendations to obtain multiple samples for analysis to improve diagnostic yield. FISH evaluation of joint fluid from this patient likewise illustrated distribution of bacteria in microcolonies (Fig. 4).

## 5 Conclusion

Molecular methods can be used to improve the diagnosis of wound and orthopedic infections. The characteristic polymicrobial nature of chronic biofilm infections is often missed with routine cultures, especially in patients who have been treated with antibiotics before samples were obtained for culture. Variability among tests supports using a variety of culture and culture-independent methods to most accurately and completely define pathogens in clinical infections. In addition, the heterogeneous nature of pathogens within biofilm communities reinforces the need to ensure broad sampling to avoid missing organisms that exist in more localized pockets within biofilms. The ongoing Danish cross-disciplinary project on PJIs holds promises for validating culture-independent diagnostics in the clinical setting with an integration of advanced clinical assessment, imaging techniques, and long-term follow-up as the reference standard.

**Acknowledgment** Thanks to our colleagues in the Danish cross-disciplinary Prosthesis: Reduction of Infection and Pain consortium. The study was supported by a grant from the Danish Agency of Science and Technology (no. 09-052174).
Thomas Bjarnsholt, Bo Jørgensen and Klaus Kirketerp-Møller are acknowledged for great collaboration in the area of chronic ulcers. The Danish Technical Research Council supported this study under the innovation consortia "BIOMED".
Martin Aasholm, Vibeke Rudkjøbing, Susanne Bielidt and Masumeh Chavoshi are thanked for their valuable technical assistance.

## References

Altschul SF, Madden TL, Schaffer AA et al (1997) Gapped BLAST and PSI-BLAST: a new generation of protein database search programs. Nucleic Acids Res 25:3389–3402

Amann RI, Krumholz L, Stahl DA (1990) Fluorescent-oligonucleotide probing of whole cells for determinative, phylogenetic, and environmental studies in microbiology. J Bacteriol 172:762–770

Andersen A, Hill KE, Stephens P et al (2007) Bacterial profiling using skin grafting, standard culture and molecular bacteriological methods. J Wound Care 16:171–175
Birch L, Dawson CE, Cornett JH, Keer JT (2001) A comparison of nucleic acid amplification techniques for the assessment of bacterial viability. Lett Appl Microbiol 33:296–301
Burmølle M, Thomsen TR, Fazli M et al (2010) Biofilms in chronic infections – a matter of opportunity – monospecies biofilms in multispecies infections. FEMS Immunol Med Microbiol 59:324–336
Davies CE, Hill KE, Wilson MJ et al (2004) Use of 16S ribosomal DNA PCR and denaturing gradient gel electrophoresis for analysis of the microfloras of healing and nonhealing chronic venous leg ulcers. J Clin Microbiol 42:3549–3557
Drancourt M, Bollet C, Carlioz A et al (2000) 16S ribosomal DNA sequence analysis of a large collection of environmental and clinical unidentifiable bacterial isolates. J Clin Microbiol 38:3623–3630
Fazli M, Bjarnsholt T, Kirketerp-Moller K, Jorgensen B, Andersen A, Krogfelt K et al (2009) Nonrandom distribution of Pseudomonas aeruginosa and Staphylococcus aureus in chronic wounds. J Clin Microbiol 47:4084–4089
Hellyer TJ, DesJardin LE, Teixeira L et al (1999) Detection of viable mycobacterium tuberculosis by reverse transcriptase-strand displacement amplification of mRNA. J Clin Microbiol 37:518–523
Horz HP, Scheer S, Huenger F, Vianna ME, Conrads G (2008) Selective isolation of bacteria DNA from human clinical specimens. J Microbiol Methods 72:98–102
Horz HP, Scheer S, Vianna ME, Conrads G (2010) New methods for selective isolation of bacterial DNA from human clinical specimens. Anaerobe 16:47–53
Kamme C, Lindberg L (1981) Aerobic and anaerobic bacteria in deep infections after total hip arthroplasty: differential diagnosis between infectious and non-infectious loosening. Clin Orthop Relat Res 154:201–207
Kirketerp-Møller K, Jensen PØ, Fazli M et al (2008) The distribution, organization and ecology of bacteria in chronic wounds. J Clin Microbiol 46:2717–2722
Kjellerup B, Thomsen TR, Nielsen JL et al (2005) Microbial diversity in biofilms from corroding heating systems. Biofouling 21:19–29
Klein PG, Juneja VK (1997) Sensitive detection of viable Listeria monocytogenes by reverse transcription-PCR. Appl Environ Microbiol 63:4441–4448
Kobayashi H, Oethinger M, Tuohy MJ et al (2009) Limiting false-positive polymerase chain reaction results: detection of DNA and mRNA to differentiate viable from dead bacteria. Diagn Microbiol Infect Dis 64:445–447
Kommedal O, Lekang K et al (2011) Characterization of polybacterial clinical samples using a set of group-specific broad-range primers targeting the 16S rRNA gene followed by DNA sequencing and RipSeq analysis. J Med Microbiol 60:927–936
Manz W, Amann R, Ludwig W, Wagner M, Schleifer KH (1992) Phylogenetic oligodeoxynucleotide probes for the major subclasses of proteobacteria: problems and solutions. Syst Appl Microbiol 15:593–600
Mikkelsen DB, Pedersen C, Højbjerg T, Schønheyder HC (2006) Culture of multiple perioperative biopsies and diagnosis of infected knee arthroplasties. APMIS 114:449–452
Muyzer G, De Waal EC, Uitterlinden AG (1993) Profiling of complex microbial populations by denaturing gradient gel electrophoresis analysis of polymerase chain reaction-amplified genes coding for 16S rRNA. Appl Environ Microbiol 59:695–700
PRIS website. https://www.knee.dk/groups/grp_login.php. Accessed Aug 2011
Sheridan GE, Masters CI, Shallcross JA, MacKey BM (1998) Detection of mRNA by reverse transcription-PCR as an indicator of viability in Escherichia coli cells. Appl Environ Microbiol 64:1313–1318
Sheridan GE, Szabo EA, Mackey BM (1999) Effect of post-treatment holding conditions on detection of tufA mRNA in ethanol-treated Escherichia coli: implications for RT-PCR-based indirect viability tests. Lett Appl Microbiol 29:375–379

Sung K, Hiett KL, Stern NJ (2005) Heat-treated Campylobacter spp. and mRNA stability as determined by reverse transcriptase-polymerase chain reaction. Foodborne Pathog Dis 2:130–137. doi:10.1089/fpd.2005.2.130

Thomsen TR, Aasholm MS, Rudkjøbing V et al (2010) The bacteriology of chronic venous leg ulcer examined by culture-independent molecular methods. Wound Repair Regen 18:38–49

Tzeneva VA, Heilig HG, van Vliet WA et al (2008) 16S rRNA targeted DGGE fingerprinting of microbial communities. Methods Mol Biol 410:335–349

Wolcott R, Gontcharova V, Sun Y, Dowd S (2009) Evaluation of the bacterial diversity among and within individual venous leg ulcers using bacterial tagencoded FLX and titanium amplicon pyrosequencing and metagenomic approaches. BMC Microbiol 9:226

Yaron S, Matthews KR (2002) A reverse transcriptase-polymerase chain reaction assay for detection of viable Escherichia coli O157:H7: investigation of specific target genes. J Appl Microbiol 92:633–640

Zhao W, Yao S, Hsing IM (2006) A microsystem compatible strategy for viable Escherichia coli detection. Biosens Bioelectron 21:1163–1170

# Improved Outcomes Via Integrated Molecular Diagnostics and Biofilm Targeted Therapeutics

**John P. Kennedy and Curtis E. Jones**

**Abstract** Molecular diagnostics are validated clinical tools available today for most all clinicians. Given the expanded microbial census they elucidate with DNA-level certainty, a new microbial reality has emerged making interpretation pivotal to advancing outcomes. Regarding chronic infection, there is little consensus for interpretation, even from traditional culture. To advance outcomes, the authors propose an escalating chronic infection protocol leveraging multiple concurrent strategies. A pivotal aspect, directing multiple legs of the protocol, is the diagnostic objectivity and accuracy of molecular diagnostics.

Using the combination of polymerase chain reaction and pyrosequencing (PCR: PSEQ) applied to chronic wounds, clinical outcomes are highlighted in a chronic infection example. Absolute healing rates for microbial DNA guided topical treatments including antibiotics and antibiofilm agents were approximately twice that of culture directed therapy ($p < 0.001$). Further, median days to closure were 177 days for culture directed therapy versus 28 days for microbial DNA guided topical treatments ($p < 0.001$).

Absent of cultivation biases and limitations, PCR:PSEQ accurately identifies all known bacteria, yeast, and fungi. While the utility of PCR:PSEQ has become relatively established for chronic wounds, the use of these methods for any chronic infection is warranted whenever the comprehensive microbial contribution is clinically relevant to interventional strategy.

## 1 Introduction

Molecular diagnostics are no longer futuristic opportunities for development. They are currently accepted and validated tools available and readily accessible for most clinicians, including reimbursement approvals. While limited outcomes-based

J.P. Kennedy (✉) • C.E. Jones
School of Pharmacy, South University, 709 Mall Boulevard, Savannah, GA 31406, USA
e-mail: jpkennedy@southuniversity.edu

G.D. Ehrlich et al. (eds.), *Culture Negative Orthopedic Biofilm Infections*,
Springer Series on Biofilms 7, DOI 10.1007/978-3-642-29554-6_4,

assessments currently exist in the literature, any implication that these are "research" methods rather than clinical tools with dramatic utility for patient care would rely upon an aversion to the validated facts. In support, the crux of diagnostics regarding microbial pathogens is *accurate* identification rather than outcomes. Outcomes are the burden of treatment strategies, the selection and efficacy of which is inextricably bound to diagnostic accuracy. Clearly the DNA-level certainty empowered by molecular methods goes unquestioned by even the most ardent proponents of traditional culture-based diagnostics. The question arises when one is faced with clinical interpretation of the results, and what a glorious opportunity that interpretation is. While debated and lacking consensus, such interpretations represent the most significant opportunity for improved outcomes of our generational watch in medical practice. To ignore or avoid such interpretations would be a dereliction of duty to advance medicine. As a whole, this text covers many diagnostic and treatment challenges linked to bacterial biofilm phenotypes. This chapter will review a definitive clinical outcomes example for chronic infection that demonstrates the power of integrating molecular diagnostics accuracy with a modern biofilm-based treatment strategy.

Before delving into the subject matter of molecular diagnostics and treatment strategies for infection, perhaps the problem, the very reason that as a clinician you are reading this text, warrants an appetizing volley, if not resolution. Firstly, great controversy surrounds the very definition of "infection." While most clinicians can ardently defend their own understanding, consensus remains elusive. For acute infection, the classic "primary signs of infection" (pain, erythema, swelling, heat, and purulent exudate) provide the preliminary diagnosis, while laboratory diagnostic methodologies are routinely employed for confirmation and identification. In fact, most acute infections have been largely and effectively addressed as a medical problem. One might argue that the failure of laboratory diagnostics to provide causality has been obscured by the significant supporting information these predatory signs for acute infection provide.

Accordingly, the clinical problem that largely remains is chronic infection, which typically follows a parasitic paradigm rather than predatory. If rigor is applied, the reader will discover that there is little consensus regarding the interpretation of cause and effect for either molecular or traditional culture-based diagnostics for chronic infection. Multiple references and clinical opinions abound, which suggest that for many tissues the presence of bacteria taken alone is not indicative of infection. Further, subjective descriptions proposed to characterize bioburden, including contamination, colonization, critical colonization, and infection, are largely academic designations without objective measures. Such designations require multiple, conditional variables that have evolved from our best understanding of acute pathology with limited relevance to chronic infection. That said, clinical signs are available in the form of secondary signs of infection (Box 1, Fig. 1; Gardner et al. 2001). These secondary signs in combination with *quantification* of bacteria would appear pivotal for chronic infection diagnosis. Subsequently, prudent reasoning would dictate that maintenance of affected tissues

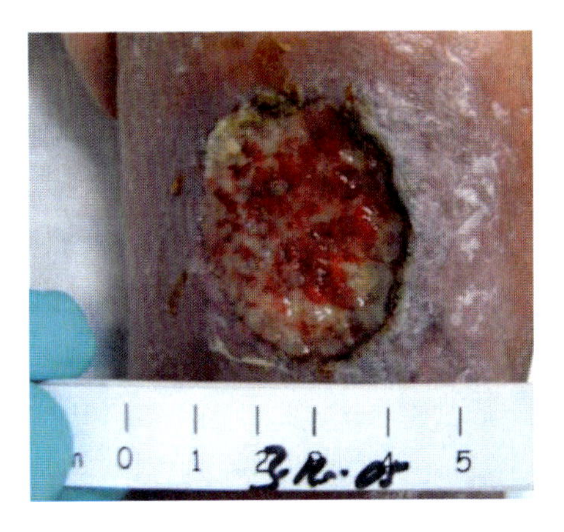


**Fig. 1** Wound showing secondary signs of infection

**Box 1. Secondary Signs of Chronic Infection [Based on Gardner et al. (2001)]**

- Serous exudate
- Tissue discoloration
- Friable tissue
- Pocketing within tissue
- Odor
- Failure to progress

as closely as possible to an unimpaired condition would most expeditiously promote the return of normal physiology and function. By way of example, under normal conditions, subcutaneous structures have no bacteria in residence of significance. Hence, the persistent inability to definitively and objectively *identify*, *quantify*, and *diagnose* the impact of bioburden hinders the standardization of treatment for many chronic infections.

Perhaps, a closing argument might be best provided by nature and the host's physical defenses. Skin and gastrointestinal cells slough regularly, predictably, and at a relatively high frequency. In our lungs, cilia assist mucus turnover. Every environmentally exposed tissue in this great design has a similar countermeasure. We are reminded of this fact two to three times a day for our dental debridement duties. However, once these physical host defenses are breached (e.g., a patient with an open wound or cystic fibrosis), tissues absent such physical defenses are exposed and at risk for biofilm maturation to outright infection (both acute and chronic). Accordingly, while not independently conclusive, any breach of host defenses should be considered partial evidence for the existence of chronic infection where significant microbes are identified.

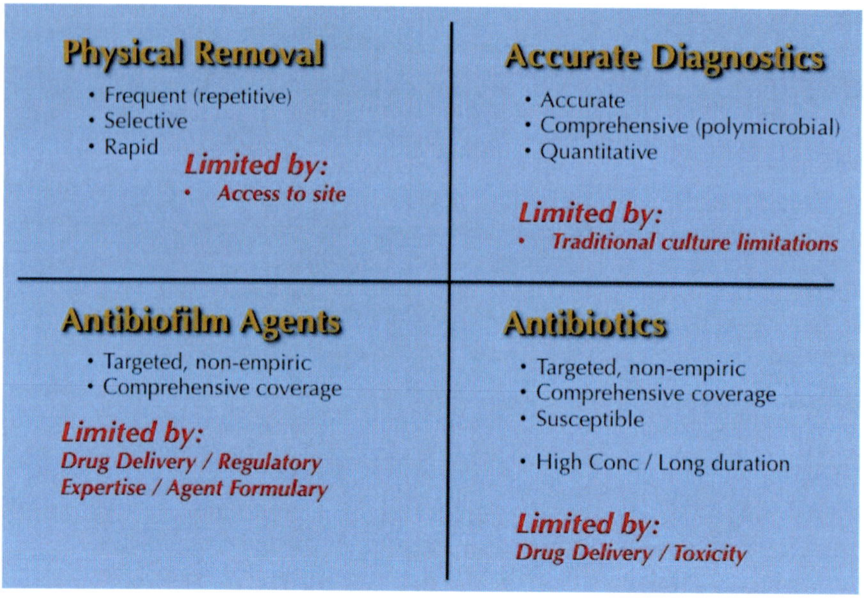

**Physical Removal**

- Frequent (repetitive)
- Selective
- Rapid

*Limited by:*

- *Access to site*

**Accurate Diagnostics**

- Accurate
- Comprehensive (polymicrobial)
- Quantitative

*Limited by:*

- *Traditional culture limitations*

**Antibiofilm Agents**

- Targeted, non-empiric
- Comprehensive coverage

*Limited by:*

*Drug Delivery / Regulatory*
*Expertise / Agent Formulary*

**Antibiotics**

- Targeted, non-empiric
- Comprehensive coverage
- Susceptible
- High Conc / Long duration

*Limited by:*

*Drug Delivery / Toxicity*

**Fig. 2** The "four-legged stool" illustration of effective biofilm management strategies [adapted from Kennedy (2011)]

## 2 Biofilm-Based Treatment Strategies for Chronic Infection

As infections become chronic, the resident bacteria naturally adopt a biofilm phenotype, and given opportunity, become mature polymicrobial communities. As opposed to acute infections, which are in alignment with Koch's postulates, chronic infections may better be described as a community of microbes acting as a singular pathology.

While perhaps not fully understanding the rationale, classic, chronic infection treatment protocols have always been comprised of inherently antibiofilm-based strategies (at least in part). For a more complete illustration of the strategic menu of options, the authors propose that any *ideal* chronic infection protocol would leverage a "four-legged stool" for chronic infection therapy. Arbitrarily, the authors have defined the four legs as (1) physical removal, (2) accurate diagnostics, (3) antibiofilm agents, and (4) antibiotics, with all components utilized in combination whenever possible to maximize outcomes (Fig. 2). This menu is preserved as ideal, regardless of etiology. In the text that follows, a brief compilation of ideal attributes within each of the four legs is provided. However, clearly the location of the affected tissue site may limit the degree of employment of any specific leg and/or attributes within.

## 2.1 Physical Removal/Debridement

The first and foundational leg is physical removal or debridement. Ideally, debridement would be employed frequently, repetitively, and selectively. Rarely is debridement absolute with regard to biofilm clearance, as residual biofilm fragments typically remain at the site even with aggressive procedures. Debridement serves as an effective broad stroke to remove the bulk of the bioburden residing at the site. As an added benefit, El-Azizi and colleagues demonstrated a marked decrease in biofilm antimicrobial resistance after biofilms were physically disrupted (El-Azizi et al. 2005). In alignment with these laboratory findings, Wolcott and colleagues demonstrated that sharp debridement opened a time-dependent optimal efficacy window for antibiotics (Wolcott et al. 2010b). Therefore, the remaining therapeutic legs serve to address such fragments, leveraging the therapeutic opportunity created via debridement. In general, without this bulk reduction of bioburden, the likelihood of success for an unaided host is progressively diminished. In the case of chronic wounds, affected tissues that cannot be debrided *regularly* are reconstituted readily. One might argue that debridement should not be classified as "foundational," as this treatment leg is not available for tissue sites that currently lack practical access (i.e., endocarditis). However, the authors feel that this designation is warranted as debridement is widely recognized as effective where it may be employed in therapy. Similarly, industrial strategies to manage biofilm universally comprise some form of physical stress or disruption as a key component to success.

## 2.2 Accurate Diagnostics

Diagnostics, the second leg, are ideally accurate, comprehensive, and quantitative. As biofilms are routinely polymicrobial, the full microbial census should be identified in order to empower a comprehensive treatment plan to target the biofilm constituents. Most polymicrobial biofilms demonstrate an astounding diversity, including yeast and fungi. Hence, accurate and comprehensive diagnostic tools are pivotal to the efficacy of the third leg and fourth leg of the paradigm (antibiofilm agents and antibiotics) in order to have confidence in a targeted and comprehensive plan.

### 2.2.1 Limitations of Traditional Culturing Techniques and Sensitivity Testing

For perspective, research microbiologists estimate that <5 % of bacteria known on this planet are capable of being identified by culture in a *research* laboratory. The number is even less for a *clinical* laboratory, to say nothing about yeast and fungi. Those disconcerting figures aside, traditional culture techniques provide limited diagnostic accuracy for biofilm-based infections, and far too often frank misrepresentation. For example, superficial wound cultures fail to consistently identify the same organisms produced from deep tissue cultures (Chakraborti et al. 2010). Simply stated, traditional cultures are selectively biased to identify the microbes their specific media was created for, often to the exclusion of dominant microbes within the specimen. Aquatic plants do not grow in desert soil, so it is for clinical culture. Further, such soils (aka media) require the correct phenotype for germination, so it is for clinical culture. For consideration, the SENTRY Antimicrobial Surveillance Program identified *Staphylococcus* as the predominant genera (~55 %) for all skin and soft tissue infections (Rennie et al. 2003). With millions of microbes given opportunity to grow on skin and soft tissues, this finding should give the reader *a priori* pause. Further, scrutiny of this result highlights a foundational bias, as <2 % of known bacteria can be grown routinely in the laboratory (Dowd and Wolcott 2010). While the ultimate conclusion from the SENTRY program identified *Staph* as the most common cause of skin and soft tissue infections, it is entirely plausible that the study more conclusively proves the bias of traditional media to grow *Staph* species.

While relevant and applicable for acute infections, sensitivity testing (susceptibility testing) offers limited value for chronic infections (e.g., infections comprising mature biofilms). It is important to understand that traditional antimicrobial sensitivity testing is conducted on planktonic bacteria grown with culture media, rather than the biofilm phenotypes attached to host tissues. It is widely understood that phenotypical differences between laboratory-grown planktonic bacteria and the clinically present biofilms of chronic infection can result in dramatic differences in antimicrobial susceptibility results (up to 1,500-fold, Lazăr and Chifiriuc 2010). The obvious question of relevancy overwhelms any perceived practical utility of traditional sensitivity testing for biofilm-based pathology. Additionally, the routine practice of limiting sensitivity testing to achievable tissue concentrations via systemic routes further limits the utility of such testing in practice for chronic infection specimens.

For all the reasons describe above and in the accompanying chapters, by default, the requirement for diagnostics to be accurate and comprehensive dictates the utilization of molecular methods for biofilm-based pathology specimens. Traditional culture-based techniques simply provide a paucity of microbial scope for such a pervasive diversity, or worse, direct clinicians to a potential solution for the wrong problem. Such results have led many clinicians to question the validity of diagnostics as a whole for specific disease states, even though the contribution of bioburden to the pathology goes unquestioned.

### 2.2.2 Molecular Diagnostics: PCR in Combination with PSEQ

For the purpose of this text, two clinically available molecular methodologies are described. Polymerase chain reaction (PCR), in combination with pyrosequencing (PSEQ) of microbial DNA, permits the identification of all known bacteria, yeast, and fungi in a specimen, regardless of the microbe's ability to grow within or upon laboratory media. More specifically, PCR provides DNA identification of pathogens based on a panel of *predetermined* primers for amplification. If the specific species resides in the specimen, it will be amplified to a detection threshold at a rate relative to its abundance. While rapid, PCR is not comprehensive as it is dependent upon the "panel" of primers selected. For this reason, PCR is largely leveraged for its speed and ability to "rule out" specific pathogens with DNA level certainty. Alternatively, PSEQ does not identify pathogens by a priori selection of markers. PSEQ is simply the sequencing of enough base pairs within microbial genetic materials to determine genera and/or species. Identification is accomplished by comparison to a library of known genera and species. Therefore, while less rapid compared to PCR (48 h vs. 1 h), PSEQ is comprehensive for all bacteria, yeast, and fungi species known. Both PCR and PSEQ results may be expressed quantitatively, depending on the laboratory and sample used for testing.

Assigning clinical relevance to the detected microbes (contamination/infection) may be difficult in some cases; however, general and specific resolutions of interpretation have been achieved and continue to expand in alignment with adoption in practice. Without primary and acute clinical signs of infection, an obvious early burden of interpretation has been the determination to treat or not to treat. Clinicians deliberate when determining if a bacterial load represents infection instead of colonization or an important pathogen rather than contamination.

Such deliberations are founded largely upon planktonic/acute paradigms rather than our more modern understanding of biofilm/chronic infections. In the case of acute, the matter is resolved readily as the primary signs of acute infection (predation) are obvious and likely the reason the specimen was originally submitted. In the case of chronic infection, secondary signs (Box 1, Fig. 1; Gardner et al. 2001) are focal points of evaluation and interpretation as such infections are more parasitic than predatory in trajectory. While accurate quantitation is available from molecular diagnostics, its role in treatment decisions is not "calibrated" and therefore remains supportive at this time. First, bacterial load thresholds in the literature are based on acute paradigms, while chronic, mature biofilms lose density over time, presumably diluted with the products of biofilm production. Second, bacterial counts referenced for acute "infection" are based on a gram of tissue or more than a cubic centimeter, which is a quantity rarely submitted for clinical diagnosis. Further, chronic infection samples, which are typically fractions of a gram, are routinely greater than the often referenced $10^5$ threshold and practically unanimously greater if a full gram is analyzed. Lastly, with regard to chronic infection, narrowly focusing on a single or predominant "pathogen" should be discouraged. While clinicians may seek a single organism to focus treatment, chronic bacterial infections are often diverse and polymicrobial, with aerobic and anaerobic bacteria co-existing as a *community* of microbes acting pathogenically as one (Dowd et al. 2008a; Sun et al. 2009). As biofilms develop, new pathogens most often join the biofilm community, further complicating the complexity of chronic infections.

**Table 1** Identified organisms from venous leg ulcers in order of diagnostic prevalence [adapted from Dowd et al. (2008b) and Wolcott et al. (2009)]

| Laboratory cultures | Molecular diagnostics |
|---|---|
| *Enterococcus* species | *Corynebacterium* species |
| *Staphylococcus* species | *Bacteroides* species |
| *Enterobacter cloacae* | *Peptoniphilus* species |
| *Pseudomonas* species | *Finegoldia* species |
| *Klebsiella* species | *Anaerococcus* species |
| *Serratia* species | *Streptococcus* species |
| *Citrobacter* species | *Serratia* species |
| *Acinetobacter* species | *Staphylococcus* species |
| | *Prevotella* species |
| No other species identified | *Peptostreptococcus* species |

Dowd and colleagues compared microbe identification in chronic wound infections with traditional culture methods and molecular diagnostics (Dowd et al. 2008b; Wolcott et al. 2009). Using the data presented for venous leg ulcers, *Enterococcus*, *Staphylococcus*, and *Enterobacter* were the top three species identified by traditional culture. Predictably, these species, as well as the others identified through culture, are known to grow most readily in routine laboratory media. However, anaerobes, bacteria with reduced metabolic activity, and bacteria requiring specialized growth environments were characteristically not identified. Conversely, PCR in combination with PSEQ of venous leg ulcers revealed a significantly broader microbial census with a vastly different order of prevalence (Table 1). Specifically, with molecular diagnostics, *Corynebacterium* was the most prevalent, *Staphylococcus* dropped down to the eighth most prevalent, while *Enterococcus* and *Enterobacter* failed to make the top ten. Significant fungi were also identified as contributors in some wounds. Such enhanced pathogen identification suggests important clinical application for chronic wounds (Table 2), a suggestion challenged in a subsequent wound study (Sect. 3).

Results from these and other molecular diagnostic evaluations of chronic wounds introduced an expanded spectrum of bacteria, clinically significant for chronic infections. In alignment, Dowd noted the frequent molecular identification of *Corynebacterium* species in wound infections, while these bacteria were historically considered to represent an insignificant nonpathogenic contaminant (Dowd and Wolcott 2010). *Corynebacterium* species and other bacteria are now understood to be pathogenic and contributors to the chronicity of biofilm infections, yet they are often not identified by culture techniques.

Molecular diagnostic techniques have revolutionized the understanding of the variety of pathogens that contribute to the pathogenicity of chronic infections. The ability to accurately and rapidly characterize the microbial census of a clinical specimen regardless of phenotype, species count, independent of cultivation bias, and with DNA level certainty supports the routine use of molecular diagnostics in infection management, both acute and chronic. In fact, these attributes are propelling clinical adoption nationwide, discipline by discipline. Molecular diagnostics will dramatically reduce, and often eliminate, ineffective, empiric treatment regimens misdirected by the inadequacies of traditional culture. The ability to

**Table 2** Comparison of microbial results obtained from wound cultures [adapted from Dowd et al. (2008b)]

| Characteristic | Laboratory culture | Molecular diagnostics |
|---|---|---|
| Identification | Only 1–5 % of all known microbes are identifiable via culture, including research settings using exhaustive methodologies | Comprehensive for known bacteria, yeast and fungi species<br>Identifies strict anaerobes routinely not identified by routine culture |
| | Limited to species that readily grow in culture media | Provides relative and absolute quantification outputs |
| Time to results | Days to week | 2–48 h |
| Diagnostic utility | As only microbes which readily grow in culture media are identified regardless of their relative contribution to the true microbial census, treatment is often "empiric" | Provides for targeted treatment selection<br>Empowers comprehensive microbial treatment strategy.<br>DNA level certainty |

improve outcomes and generate innovative therapeutic solutions, especially for the polymicrobial censuses of chronic infections, logically must begin with an accurate diagnosis.

## 2.3 *Antibiofilm Agents*

As described by the authors herein, it may be most appropriate to initially define what an antibiofilm agent is *not*. Agents with specific antibiofilm properties are not, or at least not significantly, "biocides," i.e. "true" antibiofilm agents do not produce their desired effects through cidal means (e.g., silver- or iodine-based products, which also pose significant toxicity to host cells). Nor are they selectively cidal to microbes (e.g., classical antibiotics). "True" antibiofilm agents, as defined with the context of this chapter, may help attenuate biofilms by a multitude of targets within the maturation cycle of biofilms. By way of examples, such targeted steps in the cycle include blocking biofilm attachment, impeding phenotypical biofilm metamorphosis, stimulation of phenotypical reversion back to planktonic, and inhibiting toxin production. In almost every case, antibiofilm agents are best classified as "defensive" countermeasures or agents that lower overall virulence by inhibiting aspects of biofilm maturation. All antibiofilm agents tested by the authors to date, indeed, potentiate the performance of antimicrobials (biocides and antibiotics) while they themselves lack the "offensive" potency so well known by the latter.

As true new chemical entities (e.g., antibiotics) are coming to the market at an ever decreasing rate, antibiofilm agents are a logical strategy to make the most of the antimicrobial compounds currently at our disposal, perhaps extending their clinical utility beyond present trajectories. The experience of the authors with

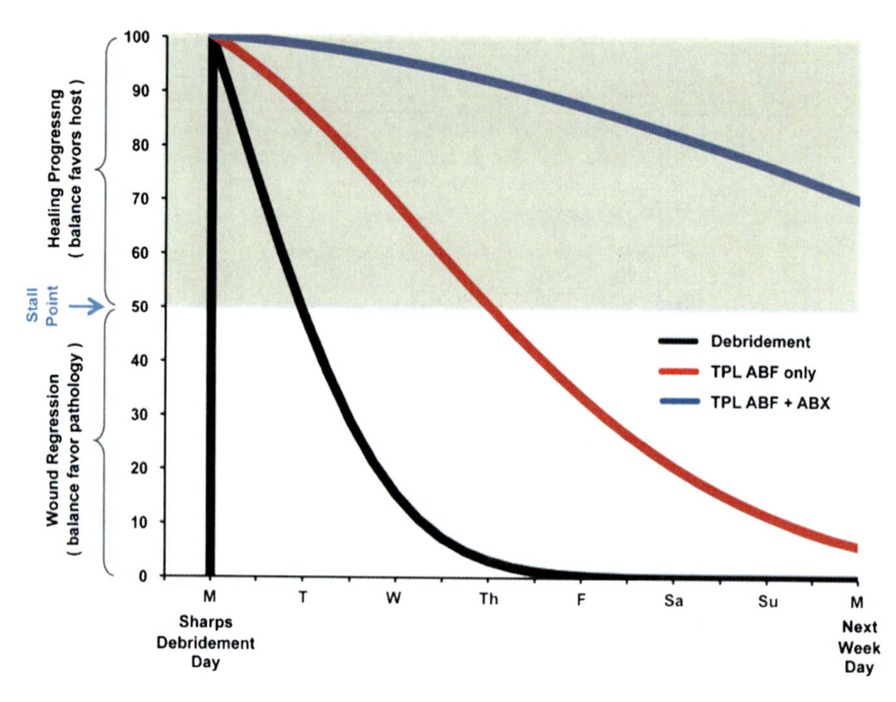


**Fig. 3** Progression/regression profile illustrating multifaceted approach to biofilm infections. In this illustration, the initial intervention is debridement, which is foundational for biofilm management across all medical disciplines. The subsequent two biofilm trajectories illustrate the advantages and relative potencies of debridement supplementation with topical ABF alone or with topical ABF + ABX in combination [adapted from Kennedy (2011)]. *ABF* antibiofilm agents, *ABX* antibiotics, *TPL* topical

these agents at this date is limited to topical and local delivery; however, systemic delivery to potentiate or extend antibiotic performance is a logical extension of utility for antibiofilm agents in future practice. Not unlike antibiotics, ideally most antibiofilm agents should be targeted, nonempiric selections that provide comprehensive coverage for the identified microbes contributing to the biofilm community.

## *2.4 Antibiotics*

For efficacy, microorganisms need to be susceptible to antibiotic selections, thereby dictating sufficient concentration and duration to eradicate infections. To date, there is no clinical sensitivity testing performed on biofilm phenotypes outside of a few research laboratories. However, clinicians may reap the knowledge gained by other disciplines that routinely treat biofilm infections, including dental medicine, otolaryngology, and ophthalmology, all of which have appreciated and leveraged

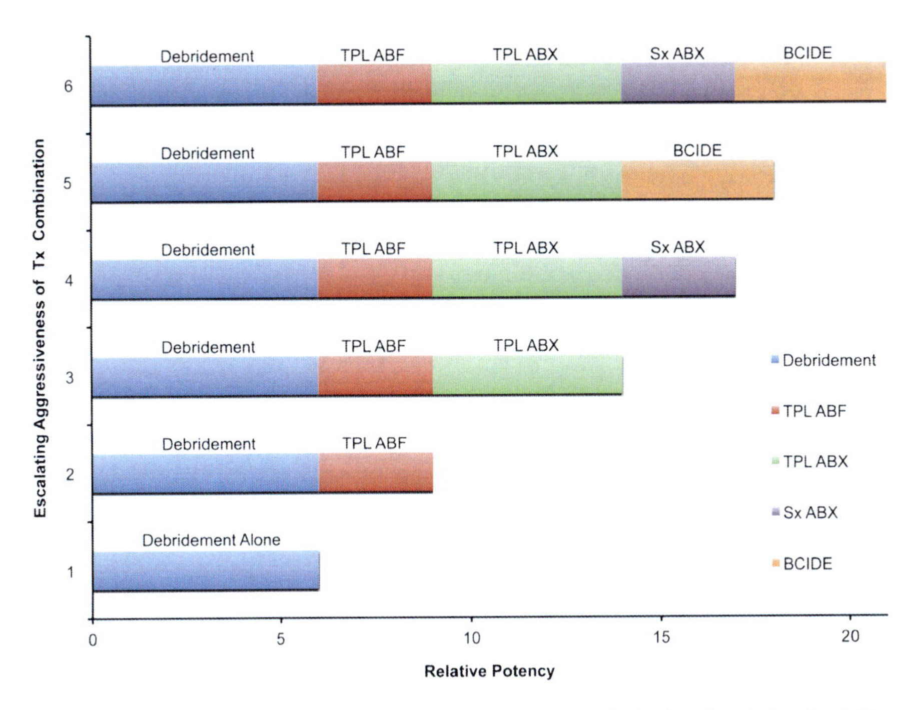


**Fig. 4** Comprehensive nature of effective biofilm management. Following foundational debridement, progressively escalating levels of interventional options are illustrated. Weekly debridement and topical ABF (defensive) combined with topical ABX (offensive) were the primary interventional levels employed in the personalized medicine study cited herein. The *addition* of Sx ABX are appropriate when deep tissues (including bone and diffuse cellulitis) are involved, but not at the exclusion of topical ABF/ABX combinations. Similarly, biocide additions offer clinical value, but are less selective (more toxic) and less targeted. Once under control, the authors' recommended prudent order for discontinuance (de-escalation) of biocides, followed by Sx ABX, and occasionally all topical interventions when wound closure is imminent [adapted from Kennedy (2011)]. *ABF* antibiofilm agents, *ABX* antibiotics, *BCIDE* biocide, *Sx* systemic, *TPL* topical

the advantages of topical and local antibiotic delivery in high concentrations over the history of each discipline. At such concentrations, far more organisms may be covered with fewer antibiotics. By way of example, eardrops used to treat chronic otitis media (clearly a biofilm infection) are over tenfold higher concentration than the highest reported resistance for *Pseudomonas*. Chronic otitis media, often polymicrobial, is clearly a biofilm-based pathology with rare treatment failures to topical therapy. A detailed accounting of the advantages of topical and local antibiotic delivery is beyond the scope of this chapter, but includes:

- Direct delivery to the targeted tissue site at high concentrations necessary to overcome biofilm phenotypical resistance
- Avoidance of pharmacokinetic limitations (absorption, metabolism, and elimination)

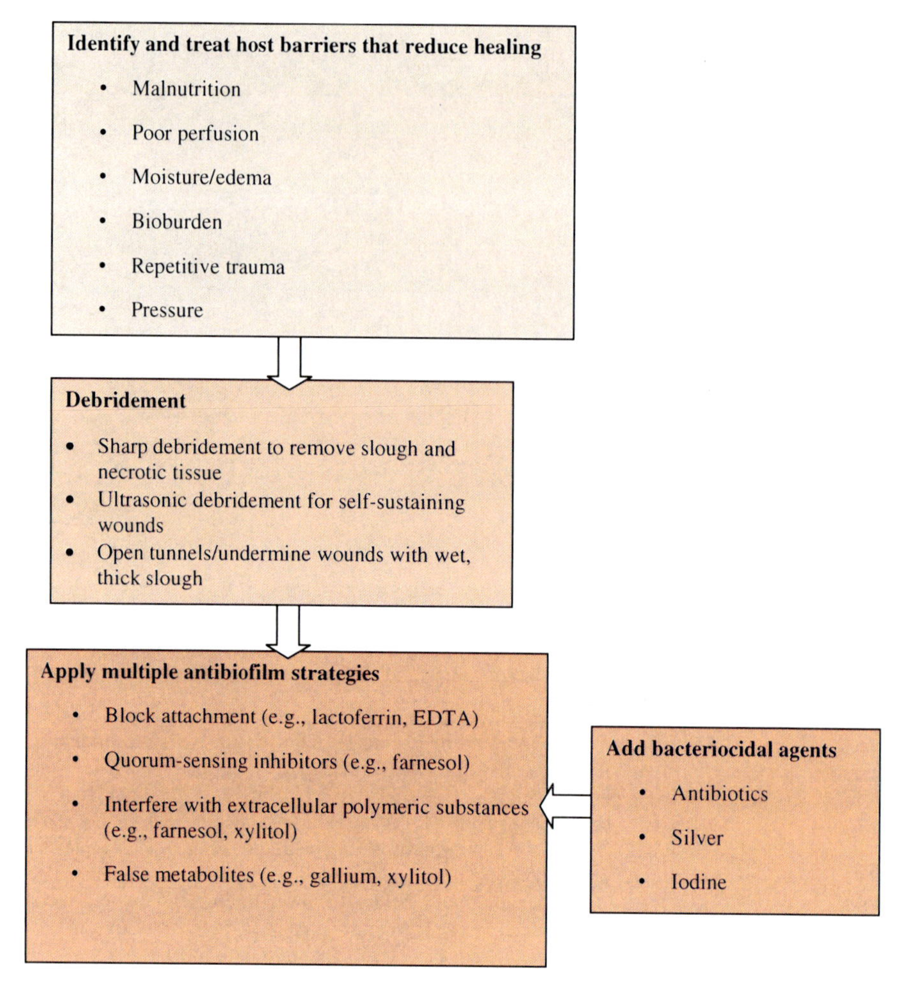


**Fig. 5** Biofilm-based wound-care algorithm [adapted from Wolcott and Rhoads (2008)]

- Better penetration of diseased tissues (largely concentration dependent)
- Lower side effects and toxicity
- Ability to leverage synergies with antibiofilm agents in topical/local combinations
- Sustained release/duration at the affected site

Successful infection management and wound healing require a combination approach to biofilm control. While debridement initially improves wound healing, debridement alone generally fails to comprehensively address bioburden, especially with regard to reconstitution of the wound bed with biofilm (Fig. 3). Without additional measures providing leverage using a comprehensive four-legged strategy, the gross benefits of debridement are progressively attenuated. A multifaceted approach that includes aggressive debridement combined with targeted antibiofilm agents and antibiotic therapy is needed to most effectively promote wound healing (Fig. 4).

Utilizing a biofilm-based wound care protocol has demonstrated improved outcomes, even for patients with significant and chronic limb ischemia (CLI) (Fig. 5; Wolcott and Rhoads 2008). For perspective, outcome comparisons with similar ischemia patients were 65 % of patient wounds *improved* using advanced wound therapy versus 77 % of patient wounds *fully closed* for the biofilm-based wound care protocol. In this biofilm-based protocol study, 75 % of patients with CLI and diabetes reached full closure by study end; further, 67 % with CLI and osteomyelitis reached full closure. The mean age of the healed group was 70.1 ± 13.2 years, with a mean transcutaneous oxygen pressure of 9 mm Hg. The authors related the clinical significance of these outcomes to the fact that limb wounds with pressures <20 mm Hg are too often managed with major amputation, with no expectation for limb salvage. This study highlights that focusing treatment on biofilm management alone is significant enough to improve treatment outcomes and salvage limbs in patients with severe wound infections.

# 3 Outcomes Example: A Fully Leveraged Four-Leg Approach

## *3.1 DNA-Identification to Define Chronic Wound Pathogens*

Wolcott and colleagues used molecular methods (PCR and PSEQ Sect. 2.2.2) to evaluate microbial census of venous leg ulcers present for >6 months (Wolcott et al. 2009). The study showed a diversity of pathogens among individual wounds, supporting the need to individually identify wound pathogens for each patient to develop targeted treatments rather than relying on empiric therapy. The study also highlighted the characteristic polymicrobial nature of these chronic wounds. In addition, a topographic variability of pathogens was noted for larger wounds, reinforcing a need for careful sampling regardless of diagnostic methodology.

## *3.2 Clinical Outcomes Driven by DNA-Guided Personalized Medicine*

Most bacteria on the human body exist as biofilms. Physical host defenses, such as skin/mucosa sloughing and cilia-assisted mucus turnover, are thought to inhibit the maturation of biofilms upon tissues and the subsequent development of outright "infection." These innate host defenses maintain the "balance" in favor of the patient. Infection occurs when these primary defenses have been breached. Further, when infections become chronic, microbial eradication is generally complicated by the

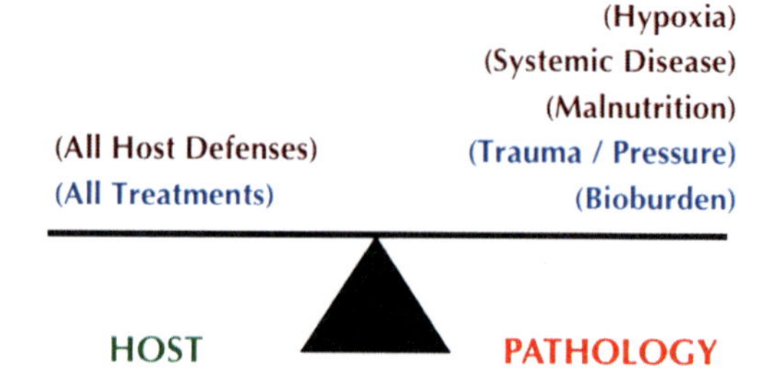


**Fig. 6** The progression/regression balance of pathology (chronic wound example) [adapted from Kennedy (2011)]

survival attributes innate to biofilm phenotypes (del Pozo and Patel 2007). As a foundational tenet of medicine, our patients heal themselves. In fact, patients will "heal" their own pathology, provided the sum of all clinical therapeutic interventions outweigh the sum of comorbidities and pathological barriers to healing, no matter how daunting. Positive outcomes simply rely upon the multiple and concurrent clinical interventions which tip the pathological balance back in favor of the host (Fig. 6).

New molecular diagnostics including PCR in combination with PSEQ are more effective tools for characterizing biofilm infections. The ability of such molecular diagnostic techniques to precisely identify pathogens in common polymicrobial biofilm infections has eliminated the need for clinicians to rely on the inadequacies of traditional culture, which is often tantamount to empiric therapy for many chronic infections.

A retrospective cohort analysis of healing rates of new, full-thickness wounds in patients treated with a biofilm-based wound care protocol at a single institution was performed before and after the introduction of molecular diagnostics (Wolcott et al. 2010a). Of the cultures conducted prior to the availability of molecular diagnostics, 23 % failed to report the presence of any bacteria. However, implementation of molecular diagnostics resulted in a significant increase in the number of bacterial species identified ($p < 0.0001$). Importantly, this increased elucidation of the true bioburden census significantly improved healing rates of patients treated with systemic antibiotics before (49 %) versus after (62 %) implementation of molecular diagnostics ($p < 0.001$) during the study period. In addition, the time to healing was reduced following implementation of the molecular diagnostic method, resulting in a 22 % reduction in time to wound closure (Box 2).

With the success of the above strategy to guide therapy, the same group sought to further define and expand the applicability of molecular diagnostics by leveraging

**Box 2. Average Reduction in Time to Heal After Implementation of Molecular Diagnostics [Based on Wolcott et al. (2010a)]**

- All new full-thickness wounds—11.8 less days to heal
- Venous leg ulcers—13.1 less days to heal
- Pressure ulcers—11.7 less days to heal
- Diabetic foot ulcers—14.2 less days to heal

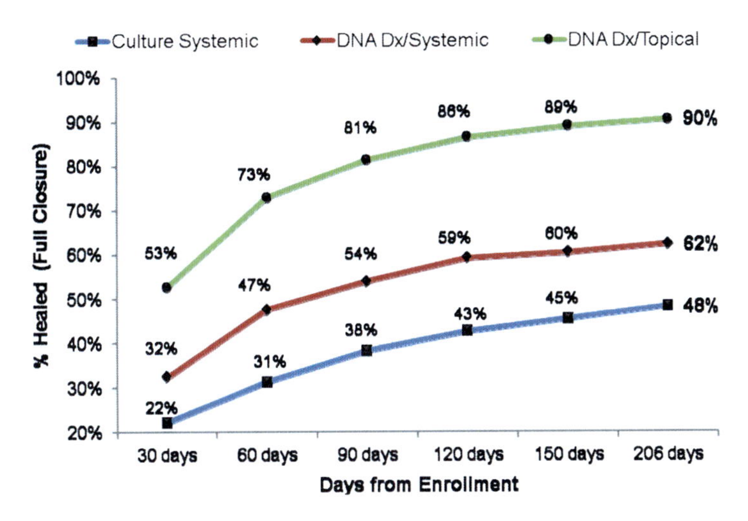


**Fig. 7** Wound healing based on molecular diagnostics. In this study, *healed* was defined as full wound closure with 100 % epithelialization [adapted from Dowd et al. (2011)]

the advantages of patient-specific topical therapies (DNA-guided personalized medicine). A retrospective cohort analysis of clinical outcomes was performed to compare healing rates in patients within three treatment groups. Each cohort was managed with a standard care utilizing multiple concurrent strategies to address bioburden based on the tenets of biofilm-based wound care (Wolcott and Rhoads 2008). Healing rates among the cohorts were compared including a standard care only group ($n = 503$) receiving systemic antimicrobial therapy guided by traditional culture-based diagnostics; treatment group 1 ($n = 479$) receiving systemic antimicrobial therapy guided by comprehensive molecular diagnostics, and treatment group 2 ($n = 396$) receiving topical antibiofilm and antibiotic therapy guided and personalized using molecular diagnostics (Dowd et al. 2011). Patients in the treatment group 2 were treated with topically administered gels that were compounded to include antibiofilm agents and tailored antibiotic coverage based on the subject's individual molecular diagnostic results. Antifungal agents were also employed topically when warranted by the corresponding diagnostic result. All patients were evaluated regularly and bioburden was addressed with frequent (weekly or bi-weekly) sharp debridement, the wound environment was maintained with appropriate advanced wound dressings, and comprehensive standard care that included reperfusion therapy, nutritional support, offloading, compression, and management of comorbidities.

Comparison of the healing rates among the cohorts revealed that the overall percentage of patients taken to complete closure over the study period increased from 48 % (244/503) in the standard care group to 62.4 % (298/479, $p < 0.001$) in treatment group 1 and 90.4 % (358/396, $p < 0.001$) in treatment group 2 (Fig. 7). In addition, the median time to closure decreased from 177 days in the standard care

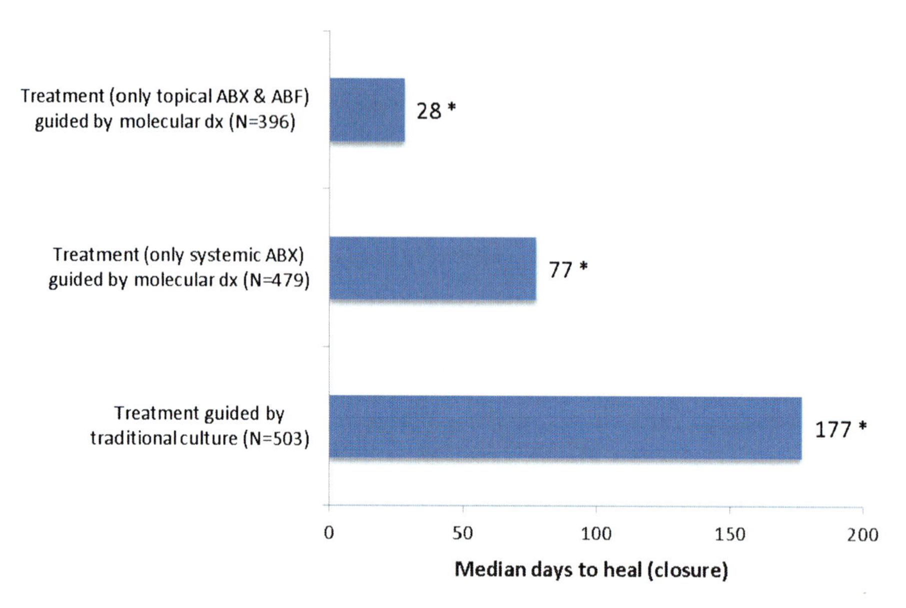


**Fig. 8** Median time to wound healing, defined as full wound closure with 100 % epithelialization. Statistical significance: $^*P < 0.001$. *ABF* antibiofilm agents, *ABX* antibiotics [adapted from Dowd et al. (2011)]

group to 77 days ($p < 0.001$) in treatment group 1 and 28 days ($p < 0.001$) in treatment group 2 (Fig. 8). While subgroup analysis by wound type showed a statistically significant difference in healing rates between the standard care group and treatment group 1 only for diabetic foot ulcers, the comparison of healing rates between the standard care group and treatment group 2, which utilized the personalized topical therapies, showed a statistically significant improvement ($p < 0.001$) in all wound types studied including diabetic foot ulcers, pressure ulcers, nonhealing surgical wounds, traumatic wounds/abscesses, and venous leg ulcers. Of note, systemic antibiotics were used in 29.4 % of patients in the standard care group (traditional culture methods) versus 50.7 % of patients in treatment group 1 in whom antibiotic selection was based on molecular diagnostic information. Predictably, the need for increased antibiotic utilization was realized with the elucidation of significantly more bacterial species requiring coverage. Patients in treatment group 2 received personalized topical therapies guided by molecular diagnostics containing, on average, three antimicrobial agents per treatment. Conversely and advantageously, this cohort received almost no systemic antibiotics (about 5 %), realizing a significant reduction to nontarget tissue and nontarget microbe exposure.

# 4 Conclusion

Biofilms are the preferred microbial phenotype for mature and persistent chronic infections. Further, they provide important barriers to effective treatment, based on structural and phenotypical changes, which offer increased antimicrobial resistance and reduced host clearance. Advanced antibiofilm strategies require accurate and objective diagnosis of these diverse polymicrobial specimens, including identification of all bacterial and fungal pathogens that may contribute to chronic disease states. Molecular diagnostics provide clinicians with the microbial reality of patient specimens, with no less than DNA level certainty.

Among the currently available molecular techniques, PCR in combination with PSEQ analysis (Sect. 2.2.2) empowers the identification of all known bacteria, yeast, and fungi in a specimen without reliance or dependence upon their ability to grow within or upon laboratory media. While rapid, PCR alone is not comprehensive as it is dependent upon the "panel" of primers selected by the user; however, PCR clearly excels for a rapid species-specific "rule-in/rule-out" diagnostic testing. Alternatively, PSEQ directly sequences the microbial DNA extracted from a clinical specimen, thereby providing identification without *a priori* assumptions regarding the microbes contained within. Though less rapid than PCR, PSEQ is comprehensive, while both methodologies may provide quantification. Traditional culture-based diagnostics simply cannot provide such objectivity, comprehensive analysis, or accuracy, especially in a chronic infection specimen subject to significantly species and phenotypic diversity. The practice of modern medicine is founded upon thorough patient evaluation, objective diagnostics, followed by comprehensive therapeutic interventions. As demonstrated herein with chronic wounds, molecular methods empower the diagnosis of many chronic infections to depart from trial and error, with no less than DNA level certainty. While the employment of molecular methods has become relatively common for chronic wounds, their employment for any chronic infection specimen is warranted whenever the microbial reality is deemed clinically relevant to therapeutic intervention.

# References


Chakraborti C, Le C, Yanofsky A (2010) Sensitivity of superficial cultures in lower extremity wounds. J Hosp Med 5(7):415–420

del Pozo JL, Patel R (2007) The challenge of treating biofilm-associated bacterial infections. Clin Pharmacol Ther 82:204–209

Dowd SE, Wolcott R (2010) Molecular diagnosis: a new era in wound care. Today's Wound Clinic 2:24–27

Dowd SE, Wolcott RD, Sun Y et al (2008a) Polymicrobial nature of chronic diabetic foot ulcer biofilm infections determined using bacterial tag encoded FLX amplicon pyrosequencing (bTEFAP). PLoS One 3:e3326

Dowd SE, Sun Y, Secor PR et al (2008b) Survey of bacterial diversity in chronic wounds using Pyrosequencing, DGGE, and full ribosome shotgun sequencing. BMC Microbiol 8:43

Dowd SE, Wolcott RD, Kennedy J, Jones C, Cox SB (2011) Molecular diagnostics and personalised medicine in wound care: assessment of outcomes. J Wound Care 20:232–239

El-Azizi M, Rao S, Kanchanapoom T, Khardori M (2005) In vitro activity of vancomycin, quinupristin/dalfopristin, and linezolid against intact and disrupted biofilms of staphylococci. Ann Clin Microbiol Antimicrob 4:2

Gardner SE, Frantz RA, Doebbeling BN (2001) The validity of the clinical signs and symptoms used to identify localized chronic wound infection. Wound Repair Regen 9:178–186

Kennedy JP (2011) Molecular enabled personalized medicine for chronic infection. Chronic orthopaedic infection conference, Center for Genomic Sciences, Allegheny-Singer Research Institute, Pittsburgh, PA, May 2011

Lazăr V, Chifiriuc MC (2010) Medical significance and new therapeutically strategies for biofilm associated infections. Roum Arch Microbiol Immunol 69:125–138

Rennie RP, Jones RN, Mutnick AH (2003) Occurrence and antimicrobial susceptibility patterns of pathogens isolated from skin and soft tissue infections: report from the SENTRY Antimicrobial Surveillance Program (United States and Canada, 2000). Diagn Microbiol Infect Dis 45:287–293

Sun Y, Smith E, Wolcott R, Dowd SE (2009) Propagation of anaerobic bacteria within an aerobic multispecies chronic wound biofilm model. J Wound Care 18:426–431

Wolcott RD, Rhoads DD (2008) A study of biofilm-based wound management in subjects with critical limb ischaemia. J Wound Care 17:145–155

Wolcott RD, Gontcharova V, Sun Y, Dowd SE (2009) Evaluation of the bacterial diversity among and within individual venous leg ulcers using bacterial tag-encoded FLX and titanium amplicon pyrosequencing and metagenomic approaches. BMC Microbiol 9:226

Wolcott RD, Cox SB, Dowd SE (2010a) Healing and healing rates of chronic wounds in the age of molecular pathogen diagnostics. J Wound Care 19:272–281

Wolcott RD, Rumbaugh KP, James G, Schultz G, Phillips P, Yang Q, Watters C, Stewart PS, Dowd SE (2010b) Biofilm maturity studies indicate sharp debridement opens a time-dependent therapeutic window. J Wound Care 19:320–328

# Immunological Methods for *Staphylococcus aureus Infection* Diagnosis and Prevention

Nathan K. Archer, J. William Costerton, Jeff G. Leid, and Mark E. Shirtliff

**Abstract** Increasing attention has been focused on understanding, diagnosing, and treating nonculturable bacterial infections. In this chapter, we explore the current immunological methods for diagnosis and prevention of recalcitrant biofilm-associated nonculturable infections in the context of the Gram-positive cocci, *Staphylococcus aureus*. In addition, we discuss immune evasion strategies of *S. aureus* in the perspective of bacteria and host, and the laboratory techniques utilized for translational research and vaccine development.

## 1 Introduction to *Staphylococcus aureus* and Biofilms

*Staphylococcus aureus* is an important biological pathogen that can result in a wide range of infections (see Box 1) and the leading cause of hospital-associated infections. Infections associated with *S. aureus* in the United States have a crude

N.K. Archer
Department of Microbial Pathogenesis, School of Dentistry, University of Maryland, Baltimore, MD, USA

Graduate Program in Life Sciences, Microbiology and Immunology Program, University of Maryland, Baltimore, MD, USA

J.W. Costerton
Center for Genomic Sciences, Allegheny-Singer Research Institute, Pittsburgh, PA, USA

J.G. Leid
Department of Biological Sciences, Northern Arizona University, Flagstaff, AZ, USA

M.E. Shirtliff (✉)
Department of Microbial Pathogenesis, School of Dentistry, University of Maryland, Baltimore, MD, USA

Department of Microbiology and Immunology, Medical School, University of Maryland, Baltimore, MD, USA
e-mail: MShirtliff@umaryland.edu

G.D. Ehrlich et al. (eds.), *Culture Negative Orthopedic Biofilm Infections*,
Springer Series on Biofilms 7, DOI 10.1007/978-3-642-29554-6_5,

**Box 1. Infections Associated with *S. aureus***
- Sepsis
- Staphylococcal scalded-skin syndrome
- Food poisoning
- Pneumonia
- Pyopneumothorax
- Empyema
- Skin infections
- Necrotizing fasciitis
- Osteomyelitis
- Endocarditis
- Keratitis
- Indwelling medical device infection
- Toxic shock syndrome

mortality rate of 25% along with hospitalizations resulting in approximately twice the length of stay, deaths, and medical costs of typical hospitalizations (Wenzel and Edmond 2001; Rubin et al. 1999). According to the National Hospital Discharge Survey, *S. aureus*-related hospital discharge diagnoses in the United States rose from almost 295,000 in 1999 to nearly 478,000 in 2005 (Klein et al. 2007). In addition, in 2005 there were also over 40,000 *S. aureus*-related deaths in the United States.

*S. aureus* may be particularly problematic because colonization is common in individuals outside of hospital environments. *S. aureus* typically lives in the anterior nares, with 80% of people intermittently or never carrying *S. aureus* and 20% serving as persistent carriers (Kluytmans et al. 1997). Carriage rates are higher among patients exposed to repeated skin puncture (e.g., dialysis patients and intravenous drug abusers) and patients with human immunodeficiency virus (HIV) infection. A survey of hemodialysis patients found that 53% were nasal carriers, with 12% of patients carrying methicillin-resistant *S. aureus* (Lederer et al. 2007). An estimated 80% of community-acquired *S. aureus* infections are caused by bacterial strains colonizing patients preoperatively (Wertheim et al. 2005). These data help validate the important role of *S. aureus* nasal colonization in subsequent community or hospital-associated infections.

The biofilm mode of growth is now recognized as a major mediator of infection, with an estimated 80% of all infections caused by biofilms (National Nosocomial Infections Surveillance 1999). Biofilms form when free-floating (planktonic) bacteria adhere to moist surfaces and develop bacterial communities with unique microstructures and survival mechanisms (Costerton et al. 1995). Common examples of *S. aureus* biofilm infections include chronic rhinosinusitis, wound infections, keratitis, indwelling medical device infections, chronic skin infections, ventilator associated pneumonia, endocarditis, and osteomyelitis (Gjodsbol et al. 2006; Hansson et al. 1995; Lew and Waldvogel 2004; Ferguson and Stolz 2005; Stephenson et al. 2010). This chapter will focus on *S. aureus* biofilm virulence and host response to infection, and relate these factors to development of immunological methods for diagnosis and infection prevention.

# 2 Immune Evasion from the Bacterial and Host Perspective

## 2.1 Immune Evasion Strategies of S. aureus

*S. aureus* uses a variety of mechanisms to evade immune responses. These include production of protein A, TSST and enterotoxin B superantigens, capsule, antigenic decoys, and leukocyte-specific toxins (Morell and Balkin 2010). In addition, *S. aureus* expresses numerous adhesins to attach to a myriad of host surfaces (Hussain et al. 1993; Resch et al. 2006). A more detailed list of mechanisms can be found in Box 2 (Foster 2005).

Another method for *S. aureus* resilience against the host response is the potential for seeding dispersal or cellular detachment from mature biofilms. Micro-colonies may detach under the direction of fluid mechanical shear forces or through a genetically programmed response that mediates the seeding dispersal process (Boyd and Chakrabarty 1994). In a similar fashion to a metastatic cancer cell, detached micro-colonies migrate from the original biofilm community to uninfected regions of the host, where they can attach and promote nascent biofilm formation. In addition, while not the case for nonmotile bacteria like *S. aureus*, seeding dispersal can also be mediated by the movement of single, motile cells from an adherent micro-colony (Sauer et al. 2002). Therefore, this advantage allows an enduring bacterial source population that is resilient against antimicrobial agents and the host immune response, while simultaneously enabling continuous shedding to encourage bacterial spread.

During an initial *S. aureus* infection, tissue-adhering and proinflammatory factors (e.g., adhesin and fibrinogen) are upregulated (Rothfork et al. 2003; Novick and Geisinger 2008). After a certain density of bacteria has been produced, a threshold is reached that results in activation of the quorum-sensing operon *agr*, which, in turn, downregulates gene expression for surface adhesins and transiently upregulates genes for capsule production, toxin secretion, and protease production (Chan et al. 2004; Boles and Horswill 2008). This allows *S. aureus* biofilms to effectively neutralize infiltrating leukocytes, such as neutrophils, that became abundant during the earlier proinflammatory phase of biofilm formation (Anwar et al. 2009).

**Box 2. *S. aureus* Features to Disrupt Normal Immune Response (Foster 2005)**

- Synthesis and cell wall anchoring of proteins that promote adhesion
- Secretion of surface proteins and polysaccharide capsule that interferes with recognition and phagocytosis
- Inhibition of neutrophil chemotaxis
- Inactivation of complement factors
- Production of leukocyte-specific toxins
- Inhibition of phagocytosis
- Secretion of cytolytic enzymes that destroy host tissues (proteases, hyaluronidase, nuclease)
- Secretion of highly immunogenic proteins that acts as an immunological decoy
- Intracellular survival
- Superantigen-dependent global T-cell activation and the resulting apoptosis
- Protein A binding to the Fc portion of host IgG resulting in both the prevention of phagocyte recognition by Fc-gamma receptors thereby preventing opsonophagocytosis and coating the microbe in Fab fragments that are not recognized as foreign

## 2.2 *Host Immune Response to Biofilm Infection*

Human leukocyte experiments reveal an unexpected interaction between leukocytes and *S. aureus* biofilm. *In vitro* experiments demonstrate that leukocytes attach to, penetrate, and produce cytokines in *S. aureus* biofilms (Leid et al. 2002). Using video microscopy, human leukocytes were shown to attach to 1-week-old *S. aureus* biofilms, initially accumulating in biofilm channels and creases, and then embedding themselves within the biofilm. Interestingly, bacteria near embedded leukocytes were not killed. Furthermore, while human leukocytes were shown to actively phagocytose planktonic *S. aureus*, they failed to phagocytose biofilm bacteria (Leid et al. 2002).

The host adaptive immune response to *S. aureus* biofilm infection was elucidated using a murine tibial implant model of osteomyelitis. Briefly, *S. aureus* biofilm-coated pins were inserted into mouse tibias. After various times post-infection to simulate the conversion from an acute infection to one that is local and chronic, the tibias along with draining lymph nodes were harvested for analysis (Prabhakara et al. 2011a). During the acute phase of infection, an inoculation dose of either $10^3$ or up to $10^5$ colony-forming units (CFU) quickly increases to $10^8$ CFU/gram bone. However, as the infection transitions to a chronic infection between days 9 and 11 post-inoculation, the bacterial levels stabilize at $10^6$ CFU/g and maintain this level for months without systemic spread and with the associated bony

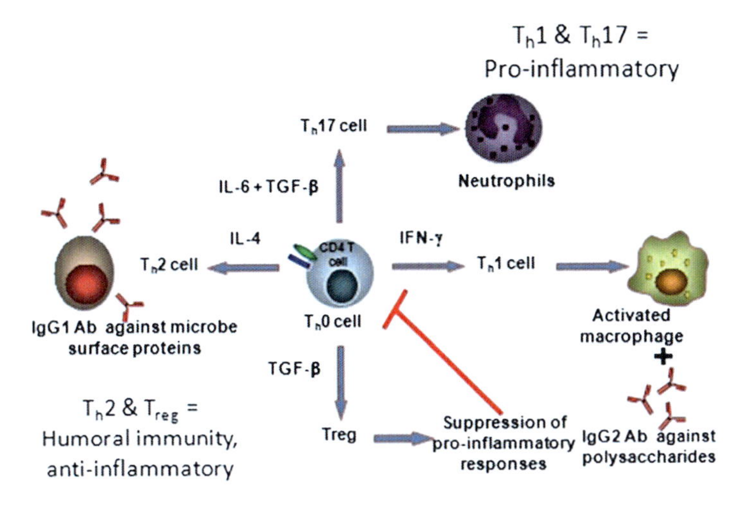


**Fig. 1** CD4+ helper response can be divided into the Th1 and Th17 proinflammatory arm on the *upper right side of the figure* and Th2 and Treg humoral response on the *lower left side*. *Ab* antibody, *Ig* immunoglobulin, *INF* interferon, *IL* interleukin, *TGF* transforming growth factor, *Th* T helper

destruction. In addition, high level vancomycin administration to which the infecting strain was sensitive was used to treat this infection following the transition to the chronic form of disease. As expected, vancomycin treatment failed to clear this infection confirming that the osteomyelitis and associated implant infection occurred through the development of an antibiotic tolerant biofilm rather than antibiotic sensitive planktonic growth.

*S. aureus* osteomyelitis upregulated proinflammatory Th1 (IL-2, IL-12p70, IFN-γ, and TNF-α) and Th17 (IL-6 and IL-17) cytokines, with the levels highest during early infection (Prabhakara et al. 2011a). These levels also corresponded with a decrease in the local anti-inflammatory Treg cell populations. Th1-dependent immunoglobulin (Ig) G2b against polysaccharides but not Th2-dependent IgG1 (recognizing microbial surface proteins) were elevated early in the infection; although this pattern reversed during later infection, when IgG2b levels decreased and IgG1 increased. This early expansion of IgG2b is ineffective since they are directed against the microbial polysaccharides and slime in the biofilm rather than attacking those cells producing these substances. Also, the IgG1 antibody levels that would be effective at clearing staphylococcal cells are not increased until over 3 weeks post-inoculation after a mature biofilm has formed and is too late to effect microbial clearance in the infection cycle. These data support that *S. aureus* biofilm infection is associated with early Th1 and Th17 inflammatory response and downregulation of Th2 and Treg responses. For a general depiction of CD4+ T helper responses please refer to Fig. 1.

Additional studies revealed that osteomyelitis infection induced expansion of CD4+ but not CD8+ T-cells, and was effectively cleared by Th2 and Treg biased BALB/c mice; however, use of Th1/Th17 biased C57BL/6J mice (like human hosts) resulted in a prolonged, chronic infection (Prabhakara et al. 2011b). This proinflammatory Th1/Th17 response would be hypothesized to support clearance of the biofilm infection through neutrophil recruitment and macrophage activation; however, a variety of virulence features help *S. aureus* evade eradication, including a resistance to phagocytosis, as mentioned previously. The differential host immune response was further highlighted in a study that assessed chronic infection rates in C57BL/6J mice receiving anti-IL-12 antibody (Ab) against the p40 antigen, thereby inhibiting Th1/Th17 responses, and BALB/c mice receiving anti-CD25 Ab (i.e., depletion of anti-inflammatory Tregs) or BALB/c STAT6 knockout (KO) mice (inhibited Th2 response) (Prabhakara et al. 2011b; Akira 1999). C57BL/6J mice that are normally unable to clear a staphylococcal biofilm infection were more effective in clearing infection when treated with anti-IL-12 Ab (inhibited Th1/Th17 responses) (Fig. 2a). Conversely, BALB/c mice that are normally able to clear a staphylococcal biofilm infection lost their ability to effectively clear *S. aureus* after receiving anti-CD25 Ab (i.e., anti-inflammatory Treg depletion) and STAT6 KO mice (inhibited anti-inflammatory Th2 response and antibody production) (Prabhakara et al. 2011b, Fig. 2b, c). These data confirm that Th1/17 responses foster chronic *S. aureus* biofilm infection, whereas Th2/Treg responses promote clearance. These immunological studies may lead to anti-biofilm therapies through appropriate immuno-modulation of the host response.

## 3 Incorporating the Biofilm Perspective into Vaccine Development and Translational Research

### *3.1 Growing* In Vitro S. aureus *Biofilms for Analysis*

*In vitro* methods for growing biofilm permit careful study of biofilm characteristics, growth patterns, and response to treatments. *S. aureus* biofilms can be cultivated in the laboratory using a flow reactor system maintained in a 37 °C incubator (Fig. 3, Brady et al. 2006). A dilute solution of CY broth (10 g casamino acids, 10 g yeast, 5 g glucose, 5.9 g NaCl, 6 mM β-glycerophosphate, and 400 mg oxacillin per liter) is pumped through silicone tubing, followed by inoculation through the tubing of medium contaminated with *S. aureus*. The bacteria are allowed to adhere to the inner surface of the tubing and grow as biofilm, after which they can be harvested for study. Biofilms can then be studied using DNA microarrays and proteomic techniques, such as two-dimensional gel electrophoresis (2DGE) to separate proteins and mass spectrometry (MS) for protein identification. This strategy of *in vitro* biofilm growth and global analysis is an effective means of identifying proteins upregulated or unique to the biofilm growth modality.

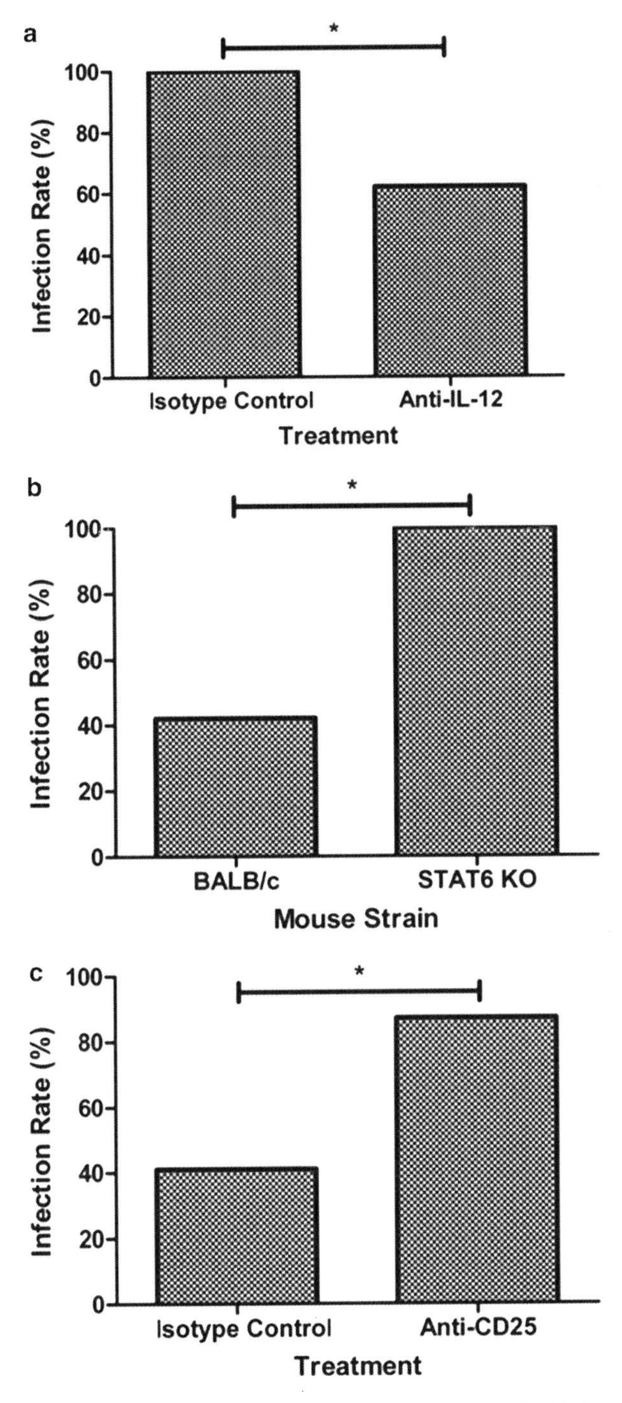


**Fig. 2** Host immune response in a murine prosthetic implant model (Prabhakara et al. 2011b). (**a**) Infection rate 21 days post-infection in isotype control Ab treated C57BL/6J mice and anti-IL-12 Ab treated C57BL/6J mice (deficient Th1/Th17 responses). $N = 6–13$ mice/group.

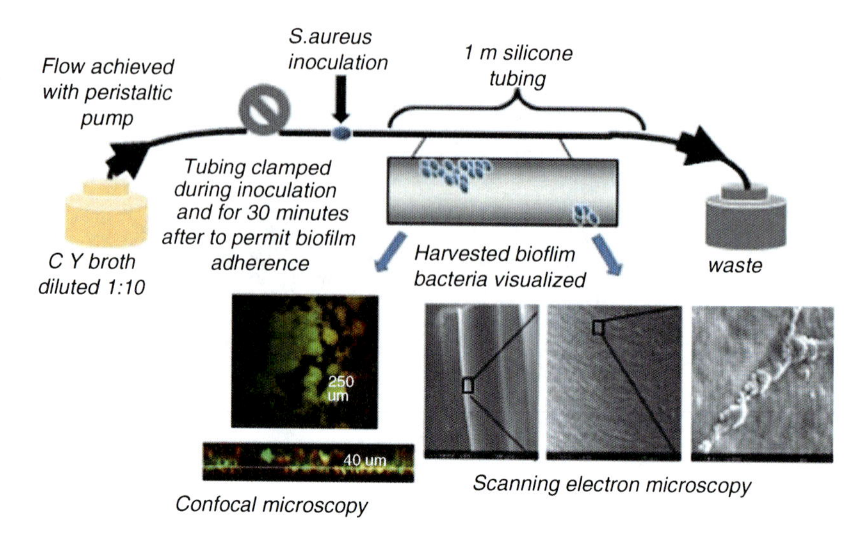


**Fig. 3** In vitro growth of *S. aureus* biofilm. CY broth contains casamino acids, yeast, glucose, NaCl, β-glycerophosphate, and oxacillin. The figure shows an enlarged image of the inside of the silicone tubing with biofilm forming and sample microscopy images of typical *S. aureus* biofilm (Brady et al. 2006)

## 3.2 Anti-Biofilm Vaccine Development

Prophylactic *S. aureus* vaccines may offer substantial reduction in morbidity and mortality related to these common infections (Stranger-Jones et al. 2006; Broughan et al. 2011). Vaccines against *S. aureus* have generally helped reduce clinical disease, but have often failed to prevent new infection (Middleton 2008). In general, antibacterial vaccines usually target one or a few virulence factors required for infection (Harro et al. 2010). *S. aureus*, however, has almost 70 virulence factors that can affect survival and resistance, with many having redundant functions (Harro et al. 2010, Table 1). Therefore, neutralization of an individual factor may have no effect on infection outcome, or can be overcome by production of another protein with similar activity. Additionally, Brady et al. discovered that antigen expression is varied throughout an individual *S. aureus* biofilm (Brady et al. 2007). Consequently, vaccines targeting specific antigens will likely eradicate only a portion of the biofilm, leaving areas without vaccinated antigen expression unaffected by the subsequent immune response. Due to the heterogeneous nature of biofilms, the variable, timed expression of antigens, and the multitude and

---

**Fig. 2** (continued) (**b**) Infection rate 21 days post-infection in BALB/c mice and BALB/c STAT6 knockout (KO) mice (deficient Th2 response). $N = 6–10$ mice/group. (**c**) Infection rate 21 days post-infection in isotype control Ab treated BALB/c mice and anti-CD25 Ab treated BALB/c mice (deficient Treg response). $N = 6–10$ mice/group. ($^{*}p < 0.05$) Ab, antibody (Prabhakara et al. 2011b)

**Table 1** Staphylococcal virulence factors

| Category | Specific factors |
|---|---|
| Adherence | Fibrinogen-binding protein |
| | %-binding proteins (A and B) |
| | Collagen binding proteins |
| | Matrix adherence protein |
| | Elastin binding protein |
| | Laminin binding protein |
| | Transferrin and lactoferrin binding protein |
| Immunoavoidance | IgG-binding protein A |
| | Capsular polysaccharides |
| | Leukocidin |
| | Biofilms |
| Host damage | ς toxin |
| | Β toxin |
| | γ toxin |
| | δ toxin |
| | Staphylococcal enterotoxins A, B |
| | C1-3, D, E, G, H, I, J, K, L, M, & O |
| | Toxins SA 1, 2, 3, 4, 5 |
| | Exfoliative toxin A (heat stable) and B (heat labile) |
| | Toxic shock syndrome toxin |
| | Phenol soluble modulins |
| Other extracellular proteins | Hyaluronidase |
| | Staphylokinase |
| | Cell free and bound coagulases |
| | Fatty acid modifying enzyme |
| | DNAse, proteases, lipases |
| | Phenol-soluble modulins |
| | Lactate dehydrogenase |
| Antimicrobial resistance | Tetracycline resistance protein |
| | β-lactamase |
| | Penicillin-binding protein 1 |
| | Quinolone resistance protein (norA) |
| | Aminoglycoside resistance protein |
| | Ethidium resistance protein (ebr) |
| | Penicillin-binding proteins 2, 2A, and 4 |
| | Cadmium resistance protein |
| | Methicillin-resistance protein |
| | Chloramphenicol acetyltransferase |
| | Kanamycin nucleotidyltransferase |
| | Bleomycin resistance protein |
| | Streptomycin resistance protein |
| | Arsenical pump membrane protein |
| | Arsenate reductase (arsC) |
| | Mercuric resistance |
| | Antiseptic resistance protein |
| | Beta-lactam-inducible penicillin binding protein |

(continued)

**Table 1** (continued)

| Category | Specific factors |
|---|---|
| | Multidrug resistance protein |
| | Glycopeptide insensitivity |
| | Vancomycin intermediate resistance |
| | Vancomycin resistance |

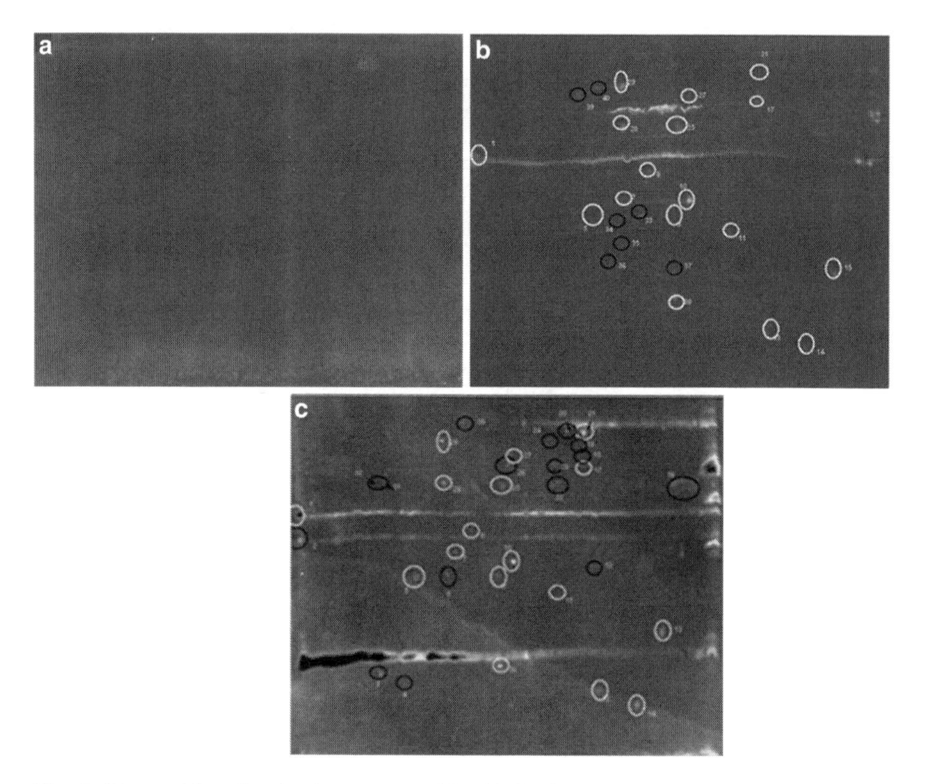


**Fig. 4** Western blots showing immunogens identified using convalescent sera with antibodies against biofilm infection in the rabbit tibial osteomyelitis model. Immunogenic proteins are *circled*. (**a**) Pre-infection sera, (**b**) 14-day sera, (**c**) 28-day sera

redundancy of virulence factors, early attempts at developing *S. aureus* vaccinations were met with limited success (Brady et al. 2011).

To combat the limitations of previous *S. aureus* vaccines, a novel technique was developed to identify biofilm antigen vaccine candidates. As mentioned above, biofilm upregulated proteins can be identified via 2DGE and MS. To identify immunogenic antigens produced by biofilms during both acute and chronic *S. aureus* infection, Brady et al. performed Western blots on 2D gels probed with serum from uninfected and staphylococcal biofilm infected rabbits (Brady et al. 2006). Figure 4 shows over 20 individual immunogens identified using this technique from *S. aureus* biofilm protein samples (Brady et al. 2006).

**Table 2** Benefits of quadrivalent biofilm-directed vaccination plus vancomycin therapy in the rabbit tibial model of *S. aureus* osteomyelitis

| Treatment group | Rabbits with clinical infection, *N* (%) | Rabbits cleared of *S. aureus*, *N* (%) |
|---|---|---|
| Control ($N = 9$) | 9 (100) | 3 (33) |
| Control plus vancomycin ($N = 9$) | 9 (100) | 5 (56) |
| Vaccine alone ($N = 9$) | 6 (66) | 2 (25) |
| Vaccine plus vancomycin ($N = 8$) | 3 (38)* | 7 (88)* |

Reduction in clinical infection and bacterial clearance were both significantly better with vaccination plus vancomycin compared with control (*$p < 0.05$) (Brady et al. 2011)

Only immunogenic antigens that were cell wall-associated and consistently biofilm upregulated were chosen for consideration in a vaccine (Brady et al. 2006). It is hypothesized that cell wall-associated antigen-specific antibodies would bind cellular *S. aureus*, resulting in enhanced phagocytosis and eradication of infection before the population of infecting microbes transitioned to a fully mature biofilm phenotype. Conversely, targeting a secreted antigen would likely have little effect on infection clearance due to the myriad and redundancy of other *S. aureus* virulence factors, and the lack of attention on cellular *S. aureus* by the immune system (Brady et al. 2011). In addition, four cell wall-associated immunogenic antigens were incorporated into a single vaccine to overcome the incomplete coverage due to biofilm protein heterogeneity of previous monovalent vaccines. Although both mono- and quadrivalent vaccines reduced radiographic infection scores, neither significantly reduced clinical signs of infection nor adequately cleared bacterial infections (Brady et al. 2011). Since the vaccine-selected proteins were upregulated in biofilms, it was postulated that the limited success was caused by lack of targeting and eradication of planktonic bacteria. Treatment became effective, however, through the combination of an antimicrobial agent treatment regimen (i.e., vancomycin) to eliminate those antibiotic-susceptible planktonic bacterial populations and the quadrivalent vaccine to target the antibiotic-resistant biofilm (Brady et al. 2011, Table 2). To produce an efficacious stand-alone *S. aureus* vaccine, future antigens would need to be present in both the planktonic and biofilm growth modalities (Harro et al. 2010). Identification of antigens by the combination of immunological techniques described above is a novel and effective means of developing vaccines against recalcitrant microorganisms, such as *S. aureus*.

### *3.3 Rapid Diagnosis Using Lateral Flow Immunoassay*

Preoperative identification of patients colonized with *S. aureus* would help identify patients for whom vaccination might be most beneficial. While traditional culture and sensitivity analyses require about 2 days for evaluation of organisms and

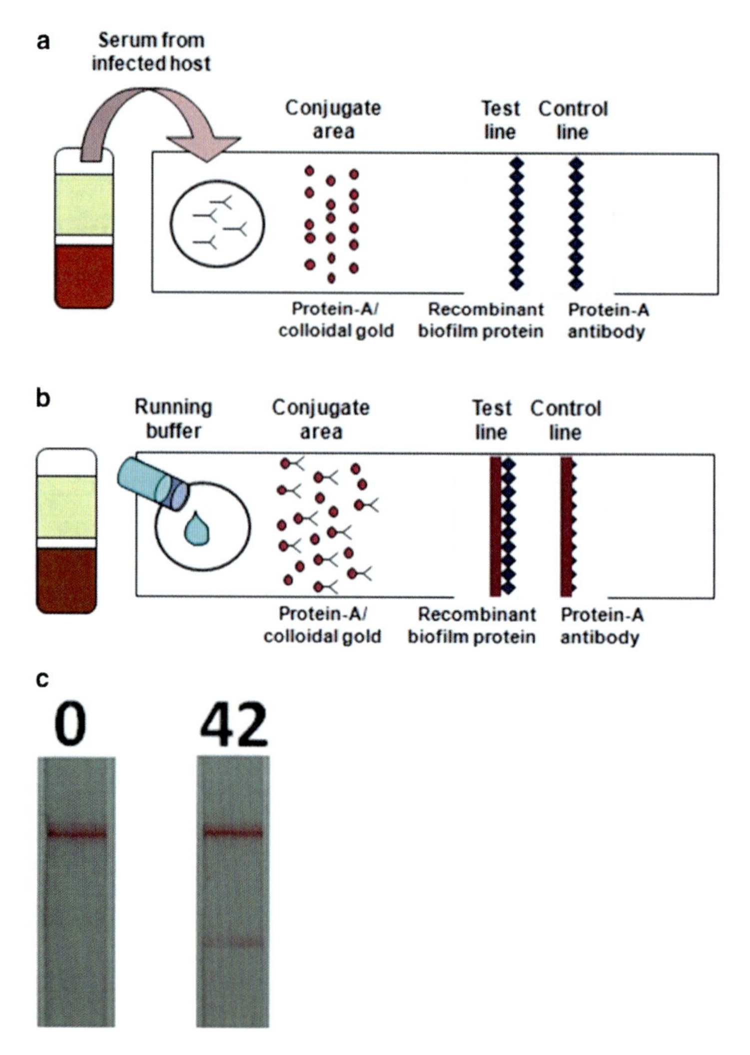


**Fig. 5** Rapid *S. aureus* screen via lateral flow immunoassay. (**a**) Immunoassay design. (**b**) Example of testing showing detection of *S. aureus*. (**c**) Sample positive test where 0 is pre-infection, and 42 are days post-infection. The *upper red bar* represents the positive control line and the *lower red bar* is the detection of antibodies against the recombinant biofilm upregulated proteins

antibiotic sensitivities, more rapid detection methods of *S. aureus* colonization would be needed to be clinically feasible. A cost analysis estimated US hospitals would prevent 935 deaths and save over $200 billion annually by performing rapid screening for *S. aureus* with subsequent decolonizing of affected patients prior to performing elective surgery (Noskin et al. 2008).

Lateral flow immunoassay has been developed as a rapid detection method for *S. aureus*. For example, cyclic probe technology with lateral flow immunoassay was able to identify methicillin-resistant *S. aureus* in 1.5 h (Fong et al. 2000). Researchers at both the University of Maryland School of Dentistry and Northern Arizona University developed an ultra-quick biofilm lateral flow immunoassay to detect host anti-biofilm antibodies against *S. aureus* biofilm-specific antigens using recombinant biofilm protein (Fig. 5). Briefly, patient serum is pulled through the test strip by capillary action and passed over the conjugate area where protein A-colloidal gold binds patient antibody (Fig. 5a, b). Serum continues to flow over the test line where specific antibodies can bind recombinant bacterial protein. A control line consists of a protein A antibody and confirms that serum has flowed completely over the test line. A red band in the test line designates a positive prognosis for *S. aureus* biofilm infection (Fig. 5c). This biofilm lateral flow immunoassay is designed to provide a very rapid diagnostic screen in the operating room or upon initial presentation, with results available within 10 min, and can detect antigen-specific antibodies within 7 days of establishment of a biofilm infection. This assay could be used to identify biofilms on implanted orthopedic prostheses and to identify patients colonized with *S. aureus* who might be candidates for vaccination therapy.

## 4 Conclusion

*S. aureus* biofilms include a wide range of virulence factors that offer unique survival advantages to reduce the effectiveness of the host immune system and traditional antimicrobial therapy. *S. aureus* infection results in an early, seemingly detrimental proinflammatory response, with increased inflammatory cytokines, neutrophils, and macrophages (Prabhakara et al. 2011a). While this initial response would be anticipated to impede infection, *S. aureus* biofilms produce toxins that damage infiltrating leukocytes and lymphocytes, thereby making the host immune mediators ineffective (Anwar et al. 2009; Thomas et al. 2007). Chronic biofilm infection can be effectively eradicated, however, through skewing of the host response towards reducing inflammation and allowing the host humoral response to proceed unimpeded. Immuno-modulation has potential as a successful immunological strategy in the elimination of recalcitrant infections.

The heterogeneous protein expression and myriad of virulence factors demonstrative of biofilm formation complicate vaccine development, often resulting in failure. A novel approach to biofilm antigen identification through use of an immunological assay and additionally combining multiple antigens, however, has resulted in development of an encouraging vaccine. When combined with antibiotic treatment to eliminate planktonic bacteria, the quadrivalent vaccine significantly reduced osteomyelitis infection in mice compared to antibiotic or vaccine treatment alone. Due to the extent of *S. aureus* persistent nasal colonization, therapeutic vaccination may be an important option in reducing the risk of subsequent nosocomial infections. Rapid detection of those patients at higher risk for *S. aureus* infection using specialized lateral flow immunoassay assessments preoperatively may help even further in attenuating nosocomial infection risk. In addition, the lateral flow immunoassay can be utilized as a diagnostic method to identify patients with recalcitrant biofilm infections, providing the knowledge needed for appropriate treatment.

Immunological methods used in diagnostic and vaccine development have proven efficacious in the context of *S. aureus* biofilm infections and disease. Further research into novel immunological methods is warranted and has potential for treatments in a wide array of bacterial infections.

## References

Akira S (1999) Functional roles of STAT family proteins: lessons from knockout mice. Stem Cells 17:138–146

Anwar S, Prince LR, Foster SJ, Whyte MK, Sabroe I (2009) The rise and rise of Staphylococcus aureus: laughing in the face of granulocytes. Clin Exp Immunol 157:216–224

Boles BR, Horswill AR (2008) Agr-mediated dispersal of Staphylococcus aureus biofilms. PLoS Pathog 4:e1000052

Boyd A, Chakrabarty AM (1994) Role of alginate lyase in cell detachment of Pseudomonas aeruginosa. Appl Environ Microbiol 60:2355–2359

Brady RA, Leid JG, Camper AK, Costerton JW, Shirtliff ME (2006) Identification of Staphylococcus aureus proteins recognized by the antibody-mediated immune response to a biofilm infection. Infect Immun 74:3415–3426

Brady RA, Leid JG, Kofonow J, Costerton JW, Shirtliff ME (2007) Immunoglobulins to surface-associated biofilm immunogens provide a novel means of visualization of methicillin-resistant Staphylococcus aureus biofilms. Appl Environ Microbiol 73:6612–6619

Brady RA, O'May GA, Leid JG et al (2011) Resolution of Staphylococcus aureus biofilm infection using vaccination and antibiotic treatment. Infect Immun 79:1797–1803

Broughan J, Anderson R, Anderson AS (2011) Strategies for and advances in the development of Staphylococcus aureus prophylactic vaccines. Expert Rev Vaccines 10:695–708

Chan WC, Coyle BJ, Williams P (2004) Virulence regulation and quorum sensing in staphylococcal infections: competitive AgrC antagonists as quorum sensing inhibitors. J Med Chem 47:4633–4641

Costerton JW, Lewandowski Z, Caldwell DE, Korber DR, Lappin-Scott HM (1995) Microbial biofilms. Annu Rev Microbiol 49:711–745

Ferguson BJ, Stolz DB (2005) Demonstration of biofilm in human bacterial chronic rhinosinusitis. Am J Rhinol 19:452–457

Fong WK, Modrusan Z, McNevin JP et al (2000) Rapid solid-phase immunoassay for detection of methicillin-resistant Staphylococcus aureus using cycling probe technology. J Clin Microbiol 38:2525–2529

Foster TJ (2005) Immune evasion by staphylococci. Nat Rev Microbiol 3:948–958

Gjodsbol K, Christensen JJ, Karlsmark T et al (2006) Multiple bacterial species reside in chronic wounds: a longitudinal study. Int Wound J 3:225–231

Hansson C, Hoborn J, Möller A, Swanbeck G (1995) The microbial flora in venous leg ulcers without clinical signs of infection. Repeated culture using a validated standardised microbiological technique. Acta Derm Venereol 75:24–30

Harro JM, Peters BM, O'May GA et al (2010) Vaccine development in Staphylococcus aureus: taking the biofilm phenotype into consideration. FEMS Immunol Med Microbiol 59:306–323
Hussain M, Wilcox MH, White PJ (1993) The slime of coagulase-negative staphylococci: biochemistry and relation to adherence. FEMS Microbiol Rev 10:191–207
Klein E, Smith DL, Laxminarayan R (2007) Hospitalizations and deaths caused by methicillin-resistant Staphylococcus aureus, United States, 1999–2005. Emerg Infect Dis 13:1840–1846
Kluytmans J, van Belkum A, Verbrugh H (1997) Nasal carriage of Staphylococcus aureus: epidemiology, underlying mechanisms, and associated risks. Clin Microbiol Rev 10:505–520
Lederer SR, Riedelsdorf G, Schiffl H (2007) Nasal carriage of methicillin resistant Staphylococcus aureus: the prevalence, patients at risk and the effect of elimination on outcomes among outclinic haemodialysis patients. Eur J Med Res 12:284–288
Leid JG, Shirtliff ME, Costerton JW, Stoodley P (2002) Human leukocytes adhere to, penetrate, and respond to Staphylococcus aureus biofilms. Infect Immun 70:6339–6345
Lew DP, Waldvogel FA (2004) Osteomyelitis. Lancet 364:369–379
Middleton JR (2008) Staphylococcus aureus antigens and challenges in vaccine development. Expert Rev Vaccines 7:805–815
Morell EA, Balkin DM (2010) Methicillin-resistant Staphylococcus aureus: a pervasive pathogen highlights the need for new antimicrobial development. Yale J Biol Med 83:223–233
National Nosocomial Infections Surveillance (1999) National Nosocomial Infections Surveillance (NNIS) System report, data summary from January 1990–May 1999, issued June 1999. Am J Infect Control 27:520–532
Noskin GA, Rubin RJ, Schentag JJ et al (2008) Budget impact analysis of rapid screening for Staphylococcus aureus colonization among patients undergoing elective surgery in US hospitals. Infect Control Hosp Epidemiol 29:16–24
Novick RP, Geisinger E (2008) Quorum sensing in staphylococci. Annu Rev Genet 42:541–564
Prabhakara R, Harro JM, Leid JG, Harris M, Shirtliff ME (2011a) Murine immune response to a chronic Staphylococcus aureus biofilm infection. Infect Immun 79:1789–1796
Prabhakara R, Harro JM, Leid JG et al (2011b) Suppression of the inflammatory immune response prevents the development of chronic biofilm infection due to methicillin resistant Staphylococcus aureus. Infect Immun 79:5010–5018
Resch A, Leicht S, Saric M et al (2006) Comparative proteome analysis of Staphylococcus aureus biofilm and planktonic cells and correlation with transcriptome profiling. Proteomics 6:1867–1877
Rothfork JM, Dessus-Babus S, Van Wamel WJ, Cheung AL, Gresham HD (2003) Fibrinogen depletion attenuates Staphylococcus aureus infection by preventing density-dependent virulence gene up-regulation. J Immunol 171:5389–5395
Rubin RJ, Harrington CA, Poon A et al (1999) The economic impact of Staphylococcus aureus infection in New York City hospitals. Emerg Infect Dis 5:9–17
Sauer K, Camper AK, Ehrlich GD, Costerton JW, Davies DG (2002) Pseudomonas aeruginosa displays multiple phenotypes during development as a biofilm. J Bacteriol 184:1140–1154
Stephenson MF, Mfuna L, Dowd SE et al (2010) Molecular characterization of the polymicrobial flora in chronic rhinosinusitis. J Otolaryngol Head Neck Surg 39:182–187
Stranger-Jones YK, Bae T, Schneewind O (2006) Vaccine assembly from surface proteins of Staphylococcus aureus. Proc Natl Acad Sci USA 103:16942–16947
Thomas D, Chou S, Dauwalder O, Lina G (2007) Diversity in Staphylococcus aureus enterotoxins. Chem Immunol Allergy 93:24–41
Wenzel RP, Edmond MB (2001) The impact of hospital-acquired bloodstream infections. Emerg Infect Dis 7:174–177
Wertheim HF, Melles DC, Vos MC et al (2005) The role of nasal carriage in Staphylococcus aureus infections. Lancet Infect Dis 5:751–762

# Diagnosing Periprosthetic Joint Infection: Cultures, Molecular Markers, and the Ibis Technology

**Javad Parvizi**

**Abstract** Lack of a standard definition or uniformly accepted diagnostic criteria offers a challenge to the management of prosthetic joint infection (PJI). Even traditional cultures fail to provide a gold standard for PJI diagnosis (Fig. 1; Parvizi 2011). Culture results depend on the sample obtained, with results often varying based on the location and method of specimen collection. Reliability of cultures depends on obtaining representative samples (including multiple tissue samples and synovial fluid) and a skilled microbiologist interpreting growth patterns. Furthermore, experts disagree on whether broth or solid media should be used for diagnostic cultures and on the number of colonies grown in culture that are needed to qualify for a positive culture. In addition, low growth rate and the frequent presence of rare organisms in cultures limit the sensitivity of culture analyses. Culture results are also limited as they merely act to confirm a preoperative diagnosis since tissue samples are obtained during a surgical procedure.

## 1 Introduction

Routinely used laboratory measures often fail to supplement determinations from gross clinical inspection. For example, Feldman and colleagues retrospectively evaluated 33 consecutive total hip or knee revision arthroplasties for which intraoperative frozen sections and long-term follow-up were available (Feldman et al. 1995). The clinical impression of the surgeon proved to be nearly as accurate as culture results and histological analyses, with a sensitivity of 0.70, specificity of 0.87, and accuracy of 0.82. These data provide concerns that

J. Parvizi (✉)
Rothman Institute of Orthopedics, Thomas Jefferson University Hospital, 925 Chestnut Street, Philadelphia, PA 19107, USA
e-mail: research@rothmaninstitute.com

G.D. Ehrlich et al. (eds.), *Culture Negative Orthopedic Biofilm Infections*,
Springer Series on Biofilms 7, DOI 10.1007/978-3-642-29554-6_6,

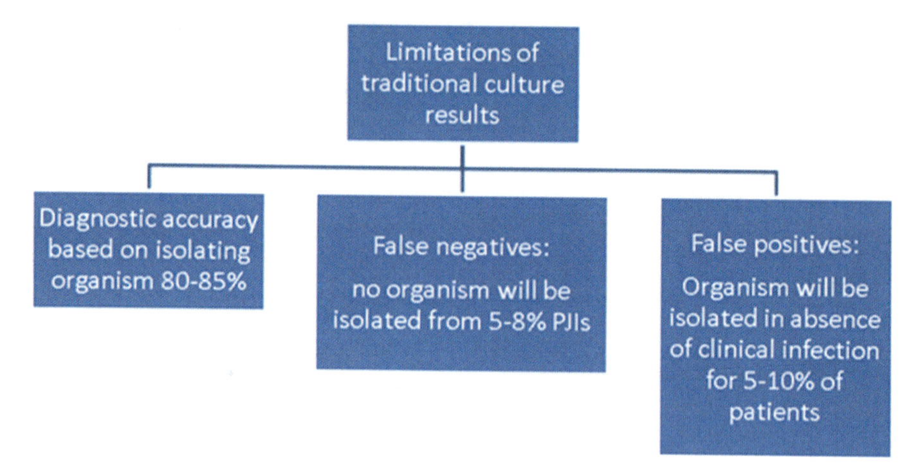


**Fig. 1** Challenges with traditional culture results for reliably diagnosing PJI [Based on Parvizi Orthopedics (2011)]

currently utilized intraoperative culture methods fail to substantially add to the diagnostic accuracy of PJIs.

## 2 Imperative to Identify PJI

Successful treatment of orthopedic patients relies on a clinically reliable diagnosis of PJI. Newly published clinical practice guidelines from the American Academy of Orthopedic Surgeons recommend withholding antibiotic therapy from patients until a diagnosis of PJI had been either reached or refuted (Della Valle et al. 2011). Confidence in diagnosing PJI is currently limited by inadequate testing options. Currently utilized tests for identifying PJI prior to or during joint removal are limited by lack of sensitivity, lack of specificity, or lack of feasibility (Table 1). Therefore, traditional laboratory measures, including culture results, fail to provide adequate direction for clinicians deciding if PJI is present.

The imperative to accurately identify PJI is further highlighted by the extensive surgery generally recommended for patients with PJI. An expert panel recently published a consensus statement, recommending two-stage revision as preferred treatment for patients with PJI (Leone et al. 2010). While success rates with two-stage revision are generally reported to be over 80% (Leone et al. 2010; Haleem et al. 2004; Bejon et al. 2010), clinicians are often hesitant to undertake this major surgery without infection confirmation with an organism isolated on laboratory testing.

**Table 1** Limitations of currently available methods for identifying PJI

| High sensitivity, low specificity | Low sensitivity, high specificity | Restricted due to expense, delay before results, or requirement of special knowledge for interpretation |
|---|---|---|
| White blood cell counts | Gram stain | Bone scans |
| Inflammatory markers (e.g., ESR and CRP) | | Polymerase chain reaction |
| Bone scans | | Cultures |
| Polymerase chain reaction | | Frozen section |

*CRP* C-reactive protein, *ESR* erythrocyte sedimentation rate

# 3 Traditional Markers of PJI

Published data from the review of a database for patients undergoing total knee arthroplasty at three academic centers ($N = 889$) found that synovial fluid analysis more accurately reflected PJI than intraoperative cultures (Parvizi et al. 2008). Optimal cutoff values >1,100 white blood cells/mcL with a >64% neutrophil differential produced a positive predictive value of 100%. Among the 116 infected patients evaluated with serum inflammatory markers, five infected patients had normal results. Positive tissue cultures had high specificity and positive predictive value; however sensitivity (30–50%) and negative predictive values (70–79%) were low. Among patients with a strong clinical suspicion for infection, intraoperative cultures produced a 10% false negative rate. Cultures were unexpectedly positive in 41 cases, of whom 29 had a single positive culture with no clinical evidence of infection or development of infection over long-term follow-up.

## *3.1 Periprosthetic Cultures*

Culture results have failed to produce a reliable gold standard for diagnosis of PJIs. Inadequacy of culture results was highlighted in a *New England Journal of Medicine* article noting that, while periprosthetic tissue culture is often used as the standard for identifying PJI, sensitivity of cultures has been reported from 65 to 94% (Zimmerli et al. 2004). Studies repeatedly show that organisms are not consistently isolated from areas that are clinically infected and positive cultures can occur in patients without clinically apparent infections (Bauer et al. 2006).

**Table 2** Median values (range) of ESR and CRP in patients with PJI or aseptic joint failure

| | Inflammatory marker | | | |
|---|---|---|---|---|
| | ESR (mm/h) | | CRP (mg/L) | |
| Joint location | Implant infection | Aseptic failure | Implant infection | Aseptic failure |
| Shoulder | 9 (1–71) | 10 (0–32) | 10 (3–40)* | 3 (3–26) |
| Spine | 48.5 (1–83)* | 10 (0–74) | 20 (3–205)* | 3 (0.5–183) |
| Hip | 30 (3–137)** | 11 (0–94) | 18 (3–288)** | 3 (0.3–141) |
| Knee | 53.5 (6–128)** | 11 (0–68) | 51 (3–444)** | 4 (0.1–174) |

Although most differences were statistically significant, high variability limits clinical usefulness
*CRP* C-reactive protein, *ESR* erythrocyte sedimentation rate [Based on Piper et al. (2010)]
$*P \leq 0.01$, $**P < 0.0001$

## 3.2 *Inflammatory Markers for Diagnosing Periprosthetic Joint Infections*

Erythrocyte sedimentation rate (ESR) and C-reactive protein (CRP) are commonly used serum inflammatory markers to identify joint infection. Reliable serum markers for PJI might help reduce unnecessary surgeries to obtain tissue cultures for a definitive diagnosis. Unfortunately, ESR and CRP often have wide variability in patients with and without infection, limiting their utility for diagnosis in individual patients. In a retrospective review of preoperative testing in 295 patients undergoing revision total knee arthroplasty, sensitivity and specificity, respectively, were reported as 0.63 and 0.55 with ESR, 0.60 and 0.63 with CRP, and 0.53 and 0.94 for intraoperative culture (Baré et al. 2006).

Serum inflammatory markers have a limited role in the preoperative PJI diagnosis due to wide variability (Table 2, Piper et al. 2010). A survey of 479 patients undergoing revision total hip arthroplasty established cutoff elevations of 31 mm/h for ESR and 20.5 mg/L for CRP as predictive for PJI (Ghanem et al. 2009a). When requiring elevations in both values, only 2.7% of cases were misdiagnosed as false negatives. Positive predictive value, however, was less robust, with 50% of patients with elevations in either measure and 26% of patients with elevations in both measures being noninfected but having a false positive diagnosis. Cutoff values could not be identified to predict infection in 109 patients undergoing two-stage total knee arthroplasty due to high test variability (Ghanem et al. 2009b). In this second study, there were no differences in inflammatory markers between patients who developed and those who did not develop PJI (Fig. 2). ESR and CRP have also been found to have poor sensitivity for the diagnosis of shoulder implant infection (Piper et al. 2010). These data question the feasibility of relying on changes in serum inflammatory markers for the diagnosis PJI.

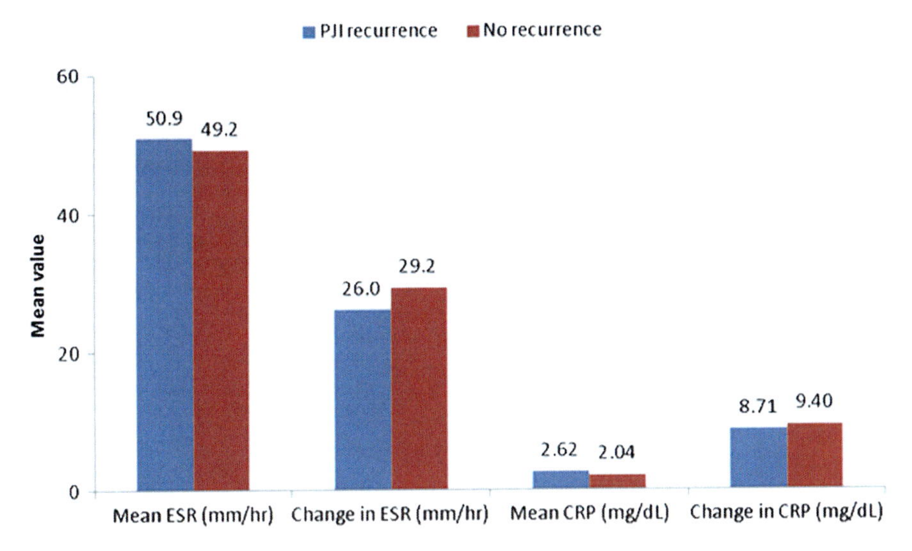


**Fig. 2** Inflammatory markers before knee prosthesis reimplantation [Based on Ghanem et al. (2009b)]. Change was calculated by subtracting values at the time of reimplantation from those at the time of resection. None of the differences between patients experiencing subsequent PJI and those with no infection recurrence was significant

## 4 New Technologies for Detecting PJI

New technologies are needed to assist clinicians in rapidly and reliably identifying PJI prior to undertaking major revision surgery. Synovial markers of inflammation may provide an effective method for assessing joint infection that may help identify patients with likely infection for whom surgery with more extensive tissue sampling might be indicated. In addition, early reports suggest that genetic evaluations via Ibis technology may provide a reliable method of identifying both PJI organisms and organism resistance, with results obtained in substantially less time than would be required for traditional culture growth. While additional evaluations of these technologies are needed, including establishing thresholds for diagnosing PJI with molecular markers, preliminary results indicate that these tests may become useful clinical tools.

### *4.1 Molecular Markers*

Synovial fluid biomarkers have been shown to be significantly elevated in PJI but not in aseptic joint failure. A prospective study measured potential inflammatory biomarkers for PJI in 31 patients undergoing revision arthroplasty for PJI and 43 undergoing similar surgery for aseptic failure (Jacovides et al. 2011).

**Table 3** Serum and synovial markers in septic and aseptic joint failure

| Biomarker | Septic patients | Aseptic patients | Between-group *P*-value |
|---|---|---|---|
| Mean ESR [mm/h (range)] | 72 (21–120) | 23 (5–73) | <0.0001 |
| Mean CRP [mg/L (range)] | 81 (<5–333) | 8.7 (5–58) | <0.0001 |
| Mean synovial leukocytes [cell/mcL (range)] | 33,545 (1,320–260,000) | 618 (0–3,299) | <0.0001 |
| Synovial neutrophils [% (range)] | 87 (37–99) | 26 (1–96) | 0.0001 |

*CRP* C-reactive protein, *ESR* erythrocyte sedimentation rate [Based on Jacovides et al. (2011)]

**Table 4** Promising synovial molecular markers of PJI

| Inflammatory protein | Sensitivity (%) | Specificity (%) | Accuracy (%) |
|---|---|---|---|
| Vascular endothelial growth factor | 77.4 | 91.5 | 85.9 |
| CRP | 87.1 | 97.7 | 93.3 |
| $A_2$-macroglobulin | 80.6 | 95.6 | 89.5 |
| Interleukin-8 | 90.3 | 97.7 | 94.7 |
| Interleukin-6 | 87.1 | 100 | 94.6 |

*CRP* C-reactive protein [Based on Jacovides et al. (2011)]

Table 3 shows that, while traditional serum and synovial inflammatory markers are significantly higher in patients with PJI compared with aseptic joint failure, the variability of each measure limits diagnostic usefulness for the individual patient. Forty-six inflammatory proteins were also measured in synovial fluid, with five proteins serving as effective molecular markers for PJI (Table 4). These data supported results from an earlier prospective study evaluating molecular markers in 14 patients with PJI and 37 patients with aseptic failure (Deirmengian et al. 2010). In that study, 12 synovial biomarkers were substantially higher in patients with PJI, with interleukin (IL)-1 and IL-6 correctly classifying all patients with sensitivity, specificity, and accuracy values of 100%. Synovial IL-6 was significantly more accurate for diagnosing PJI than the serum inflammatory markers ESR ($P < 0.001$) and CRP ($P = 0.003$) or synovial leukocytes ($P = 0.013$) or percentage of segmented neutrophils ($P = 0.001$).

Another promising molecular marker for PJI is the measurement of CRP in the synovial fluid. Based on preliminary study, it appears that CRP in the synovial fluid carries a much higher specificity than the CRP in serum (Parvizi et al. 2012).

## 4.2 *Ibis Technology*

The Ibis technology is based on PCR amplification and mass spectrometric analysis of specific segments of 16S ribosomal RNA (rRNA) genes and multiple other genes in bacteria. Extraction and analysis of bacterial nucleic acids may be used to identify molecular markers of PJI (Parvizi et al. 2012). Bacterial sequences are amplified and their molecular masses are compared with a stored database of molecular masses for species-specific organism identification (Sampath et al. 2007). The Ibis technology provides rapid identification of bacteria and determination of resistance within 6 h. Sensitivity is similar for both planktonic and biofilm infections.

The Ibis technology can also identify resistant organisms by identifying antibiotic resistant genes, such as *mecA*, which codes for oxacillin resistance. Stoodley and colleagues used the Ibis technology to identify previously undetected methicillin-resistant *S. aureus* in a patient undergoing ankle arthroplasty (Stoodley et al. 2011). Reverse transcriptase-PCR was used to identify metabolically active bacterial mRNA, with the Ibis biosensor added to provide rapid and accurate detection of bacteria and antibiotic resistance.

## 4.3 Molecular Markers and Ibis Technology

During a 10-month period in 2009, data were prospectively collected from consecutive septic and aseptic knee and hip revisions and primary total knee arthroplasty procedures performed by six surgeons at a single institute (unpublished data). Primary total knee arthroplasty patients served as a comparison control group. Five intraoperative tissue samples from representative areas and synovial fluid were obtained from each patient. This was an observational study so study data did not influence clinical treatment. Synovial fluid was aspirated before arthrotomy, with the syringe needle changed to control for contamination. Samples were placed in a cryotube and transferred to a liquid nitrogen freezer (−140 °F) and subsequently evaluated using Ibis analyses. Patients were evaluated with traditional measures (e.g., inflammatory markers and cultures), as well as molecular markers and Ibis identification of infecting organisms and resistance via the *mecA* gene. Molecular markers included a broad range of 49 potential compounds identified in synovial fluid, including cytokines, adhesion molecules, growth factors, acute phase reactants, complement cascade proteins, metalloproteinases compounds, lysis/destruction proteins, and others. The most interesting marker was synovial CRP.

PJI diagnosis was established using preoperative physician judgment and standardized criteria (Table 5). A total of 73 patients were included in this sample undergoing knee or hip revision, with 12 primary total knee arthroplasty controls. Based on the diagnostic criteria used, 28 were classified as septic revisions and 41 as aseptic, with some disagreement between the surgeon's clinical impression and standardized criterion results (Table 6). Four cases were indeterminate, based on lack of sufficient serology results. Demographics were similar between infected and uninfected patients, respectively, for age (80.9 vs. 86.5 years, $P = 0.407$), male gender (53.6 vs. 44.4%, $P = 0.08$), and revised joint (64.3% knee vs. 64.4% knee, $P = 1.0$). Body mass index showed a trend toward being higher among uninfected patients (30.8 kg/m$^2$ in infected vs. 34.2 kg/m$^2$ in uninfected patients, $P = 0.077$).

**Table 5** Criteria for diagnosing PJI adopted by Rothman Institute of Orthopedics

| PJI is diagnosed when either of the following criteria is met | |
|---|---|
| At least one of the following | At least three of the following serology results |
| Positive culture | ESR >30 mm/h |
| Intraoperative purulence | CRP >10 mg/L |
| Draining sinus tract | White blood cell count >1,760 cells/mcL OR ≥10,700 cells/mcL in acute postoperative synovial fluid |
| | Percent neutrophils >73% OR ≥89% in acute postoperative synovial fluid |

*CRP* C-reactive protein, *ESR* erythrocyte sedimentation rate

**Table 6** Diagnosis using surgeon judgment vs. standard criteria described in Table 5

| | | Surgeon judgment | | |
|---|---|---|---|---|
| | | Septic | Aseptic | Total |
| Standardized diagnostic criteria (see Table 5) | Septic | 24 | 4 | 28 |
| | Aseptic | 2 | 39 | 41 |
| | Indeterminate | 0 | 4 | 4 |
| Total | | 26 | 47 | 73 |

Conventional culture was positive in 21 of 28 patients meeting standardized infection criteria. Synovial CRP suggested infection in 32 patients, two of whom were later determined not to have an infection. Discordance for synovial CRP for identifying PJI was about 5%.

Ibis technology identified organisms that had been missed on traditional culture (Fig. 3). Figure 3a shows that Ibis effectively identified all of the significant organisms that were subsequently grown in culture and was more effective in determining methicillin resistance. Among the seven patients diagnosed with infection for whom culture results were negative (Fig. 3b), *Staphylococcus* or *Streptococcus* species were identified in four cases with Ibis and a clinically important fungal infection was also identified by Ibis.

# 5 Future Directions for Diagnosing PJI

Molecular markers and Ibis technology may provide useful tools for identifying PJI in patients prior to proceeding with surgery. Molecular markers, such as IL-6 and synovial CRP, may offer useful measures for PJI, although additional studies are needed to establish reliable thresholds for infection. Ibis technology may provide a rapid and accurate determination of infection and antibiotic resistance that might help direct subsequent surgical and medical treatment. Ibis may be able to provide guidance for clinicians with culture-negative patients when clinical suspicion is otherwise high for infection. Ibis may additionally be used to confirm culture results and establish antibiotic resistance in patients for whom cultures have isolated specific organisms.

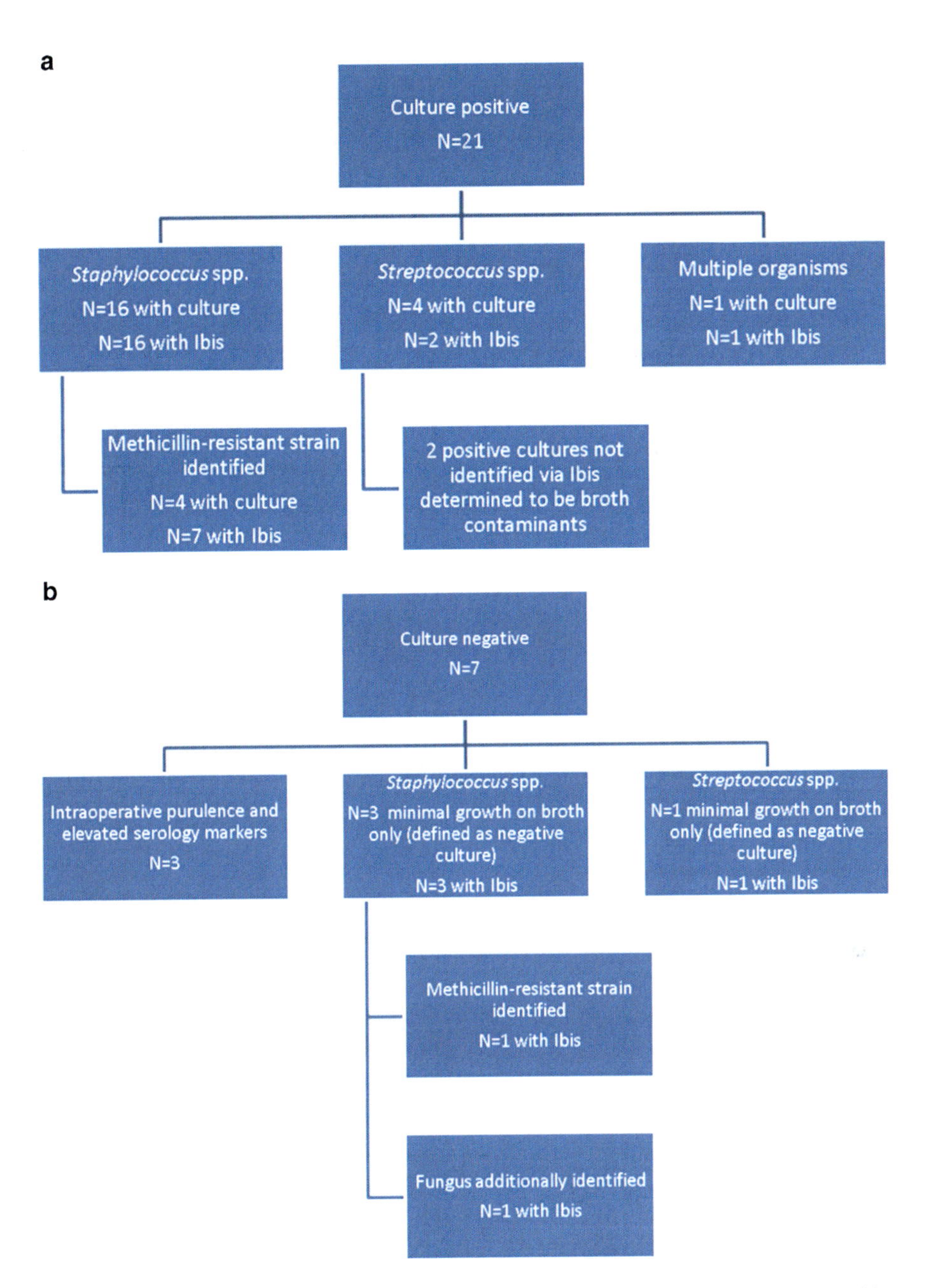


**Fig. 3** Culture vs. Ibis results in patients who met standard infection criteria ($N = 28$). (**a**) Infected with positive culture results. (**b**) Infected with negative culture results

# References


Baré J, MacDonald SJ, Bourne RB (2006) Preoperative evaluations in revision total knee arthroplasty. Clin Orthop Relat Res 446:40–44

Bauer TW, Parvizi J, Kobayashi N, Krebs V (2006) Diagnosis of periprosthetic infection. J Bone Joint Surg Am 88:869–882

Bejon P, Berendt A, Atkins BL et al (2010) Two-stage revision for prosthetic joint infection: predictors of outcome and the role of reimplantation microbiology. J Antimicrob Chemother 65:569–575

Deirmengian C, Hallab N, Tarabishy A et al (2010) Synovial fluid biomarkers for periprosthetic infection. Clin Orthop Relat Res 468:2017–2023

Della Valle C, Parvizi J, Bauer TW et al (2011) American Academy of Orthopaedic Surgeons clinical practice guideline on: the diagnosis of periprosthetic joint infections of the hip and knee. J Bone Joint Surg Am 93:1355–1357

Feldman DS, Lonner JH, Desai P, Zuckerman JD (1995) The role of intraoperative frozen sections in revision total joint arthroplasty. J Bone Joint Surg Am 77:1807–1813

Ghanem E, Antoci V, Pulido L et al (2009a) The use of receiver operating characteristics analysis in determining erythrocyte sedimentation rate and C-reactive protein levels in diagnosing periprosthetic infection prior to revision total hip arthroplasty. Int J Infect Dis 13:e444–e449

Ghanem E, Azzam K, Seeley M, Joshi A, Parvizi J (2009b) Staged revision for knee arthroplasty infection. What is the role of serological tests before reimplantation? Clin Orthop Relat Res 467:1699–1705

Haleem AA, Berry DJ, Hanssen AD (2004) Mid-term to long-term followup of two-stage reimplantation for infected total knee arthroplasty. Clin Orthop Relat Res 428:35–39

Jacovides CL, Parvizi J, Adeli B, Jung KA (2011) Molecular markers for diagnosis of periprosthetic joint infection. J Arthroplasty 26(6 Suppl):99–103.e1

Leone S, Borrè S, Monforte A et al (2010) Consensus document on controversial issues in the diagnosis and treatment of prosthetic joint infections. Int J Infect Dis 14(Suppl 4):S67–S77

Parvizi J (2011) Periprosthetic joint infection. Orthopedics 34:448–449

Parvizi J, Ghanem E, Sharkey P et al (2008) Diagnosis of infected total knee. Findings of a multicenter database. Clin Orthop Relat Res 466(11):2628–2633

Parvizi J, Jacovides C, Adeli B, Jung KA, Hozack WJ (2012) Mark B. Coventry Award: synovial C-reactive protein: a prospective evaluation of a molecular marker for periprosthetic knee joint infection. Clin Orthop Relat Res 470:54–60

Piper KE, Fernandez-Sampedro M, Steckelberg KE et al (2010) C-reactive protein, erythrocyte sedimentation rate and orthopedic implant infection. PLoS One 5:e9358

Sampath R, Hall TA, Massire C et al (2007) Rapid identification of emerging infectious agents using PCR and electrospray ionization mass spectrometry. Ann N Y Acad Sci 1102:109–120

Stoodley P, Conti SF, DeMeo PJ et al (2011) Characterization of a mixed MRSA/MRSE biofilm in an explanted total ankle arthroplasty. FEMS Immunol Med Microbiol 62:66–74

Zimmerli W, Trampuz A, Ochsner PE (2004) Prosthetic-joint infection. N Engl J Med 351:1645–1654

# Infections Associated with Severe Open Tibial Fractures

**Robert V. O'Toole**

**Abstract** Severe open tibial fractures are frequently complicated by infection. There is some data that external fixation of these fractures results in fewer and less serious infections than internal fixation. However, there is insufficient data in this regard to justify such a recommendation at this juncture. Therefore, there is a pressing need for a multi-center, prospective, randomized clinical trial to ascertain if there are statistically significant benefits of one fixation type versus the other for tibial fracture. Towards that end we have developed the FixIt Trial which aims to enroll 300 patients with type IIIB or "severe" type IIIA tibial fractures from 23 major civilian and military trauma centers for randomization of treatment and clinical follow-up for one year.

## 1 Introduction

The most severe open tibial fractures are often associated with poor outcome and are frequently complicated by infection. Unlike other infections presented in this text, infection after high-energy fracture is typically culture positive in both open and closed injuries, even in patients who have already received antibiotics. For example, in one recent prospective infection trial of fixation of high-energy fractures, cultures were positive 85% of the time that infection was identified (Stall et al. 2010). Open fractures may be classified based on injury severity (Table 1, Gustilo et al. 1984). The more severe types are often caused by high-energy trauma, such as motorcycle or motor vehicle accidents, pedestrian accidents, and gunshot wounds (Gustilo et al. 1984). Figure 1 shows an example of a Type IIIB tibial shaft fracture after flap coverage.

R.V. O'Toole (✉)
Department of Orthopaedics, R Adams Cowley Shock Trauma Center, University of Maryland School of Medicine, Baltimore, MD 21201, USA
e-mail: rvo3@yahoo.com

G.D. Ehrlich et al. (eds.), *Culture Negative Orthopedic Biofilm Infections*,
Springer Series on Biofilms 7, DOI 10.1007/978-3-642-29554-6_7,

**Table 1** Classification of open fractures [Based on Gustilo et al. (1984)]

| Severity category | Description |
|---|---|
| Type I | Clean wound <1 cm with a simple fracture pattern and no crush injury |
| Type II | Laceration >1 cm with no crush injury. Fracture pattern may be more complex |
| Type III | Open segmental fracture or single fracture accompanied by extensive soft-tissue damage |
| IIIA | Adequate fracture coverage with soft tissue |
| IIIB | Requires flap coverage. Extensive soft-tissue injury with periosteal stripping and bone exposure. Marked contamination is likely |
| IIIC | Open fracture with vascular injury that requires repair for limb salvage |

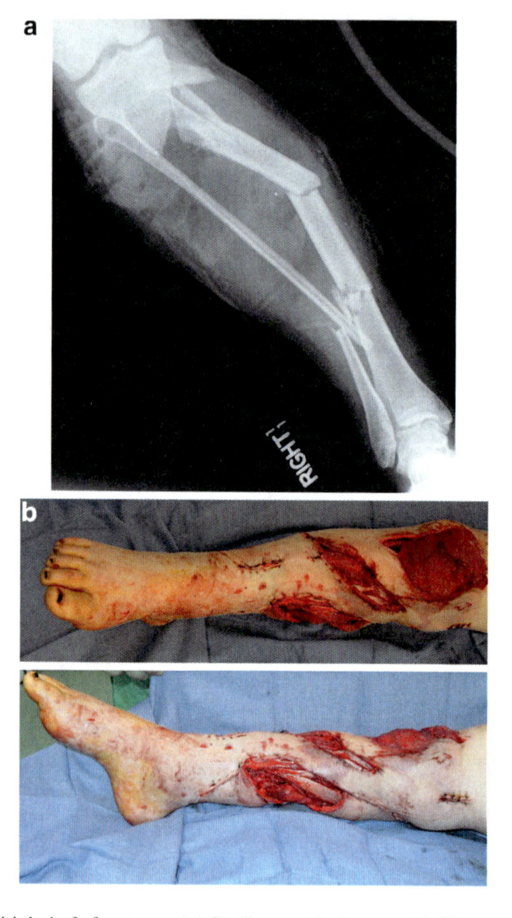


**Fig. 1** Type IIIB tibial shaft fractures. (**a**) Radiographic image. (**b**) Photograph of patient from whom above radiograph was obtained. Photograph was obtained after multiple rotation flaps covered open wounds so that bone was no longer exposed

**Table 2** 24-Month outcome in patients with severe open tibia fractures [Based on Bosse et al. (2002)]

| | Tibia fracture severity | | |
|---|---|---|---|
| Outcome (%) | IIIA ($N = 27$) | IIIB ($N = 135$) | IIIC ($N = 11$) |
| Late amputation or stump revision | 3.7 | 5.2 | 9.1 |
| Non-healing fracture | 4.4 | 13.0 | 12.5 |
| Non-healing soft tissue | 0 | 3.5 | 0 |
| Additional surgery required | 7.7 | 20.5 | 30.0 |
| ≥1 rehospitalization | 59.3 | 57.0 | 45.5 |

A large, multicenter, prospective, observational study published in 2002 reported long-term outcome with high-energy lower extremity trauma that included both severe IIIA fractures as well as type IIIB fractures (the LEAP study, Bosse et al. 2002). Over half of the patients required rehospitalization. Osteomyelitis or other infection, respectively, resulted in rehospitalization for 18.5 and 14.8% of patients with Type IIIA fracture, 11.1 and 14.1% with Type IIIB, and 27.3 and 27.3% with Type IIIC (Table 2). Infection is also a common complication in combat-related severe tibial fractures, as these injuries are typically more severe and amputation more likely compared with civilian injuries (Burns et al. 2010; Doucet et al. 2011).

## 2 Mechanism of Infection in Severe Open Tibial Fractures

A variety of factors may increase infection risk in patients experiencing Type III tibial fractures. These include local damage to blood supply from the poly-traumatized limb, poor physiologic reserve from a polytraumatized host, and a generally poor underlying host immune system that may be more common in civilian trauma patients than the general population. In addition, the fracture has historically been stabilized by a metal device (an intramedullary nail or plate), which may provide important surfaces for biofilm attachment and growth. Biofilm grows readily on a wide range of biomaterial implants (Gottenbos et al. 2000). For example, in one study, *Staphylococcus epidermidis* adhered to stainless steel within 25 min of exposure, reaching maximum growth volume within 4.3 h (van der Borden et al. 2004).

## 3 Internal vs. External Fixation

Stabilization of severe fractures may be achieved through internal or external fixation, each of which offers treatment advantages and disadvantages (Table 3). Internal fixation may be achieved through the use of intramedullary nails (Fig. 2) or plates and screws. Both of these methods insert metal at the fracture site, which may

**Table 3** Comparison of internal and external fixation for severe open tibial shaft fractures

| Type of fixation | Advantages | Disadvantages |
|---|---|---|
| Internal | Orthopedists are most familiar with internal fixation<br>All hardware is contained within the skin which may be preferable to patients | Metal is present at the fracture site, providing a potentially undesirable nidus for biofilm growth |
| External | Lack of metal at the fracture site could reduce deep infections | Orthopedists are less familiar with modern ring external fixation devices and techniques<br>Care required by patient and family is more complex<br>Pin-tract infection is common<br>Fixation time is long |

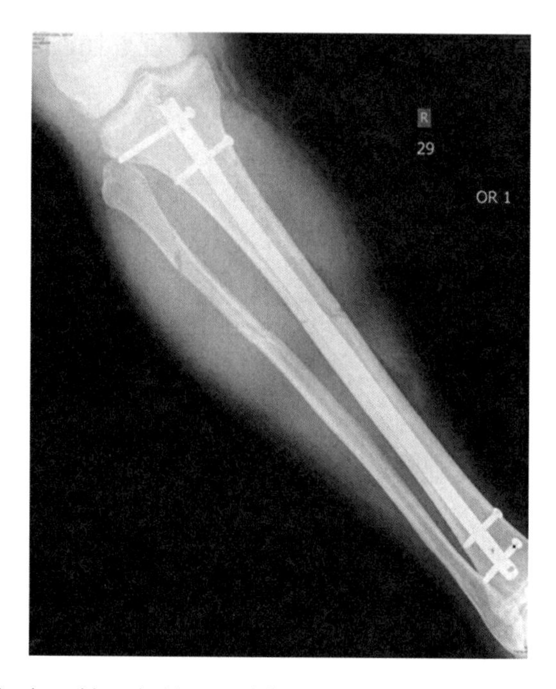


**Fig. 2** Internal fixation with typical intramedullary nail of tibial shaft fracture

serve as a nidus for biofilm growth. External fixation is currently achieved through the use of ring fixators (Fig. 3) that offer a potentially important advantage that there is no metal at the fracture site, which may reduce infection. Modern ring fixators may have overcome the problem of malunion in previous external fixators through improved mechanics and sequential computer-controlled correction to facilitate bone alignment in the clinic. Pin-site infection is a frequent complication of external fixation, as a tract is formed from surface bacteria to deeper tissues;

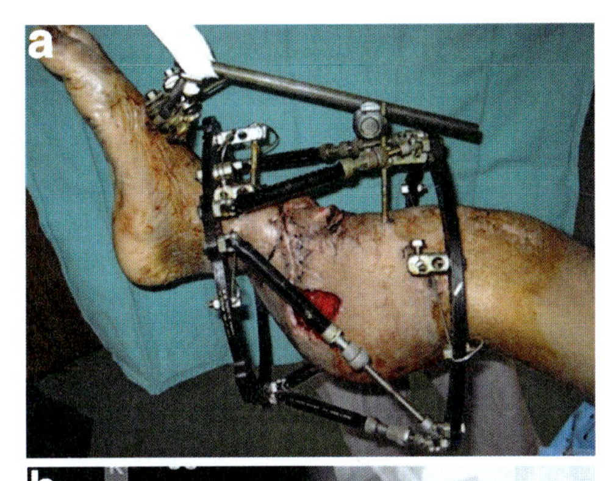


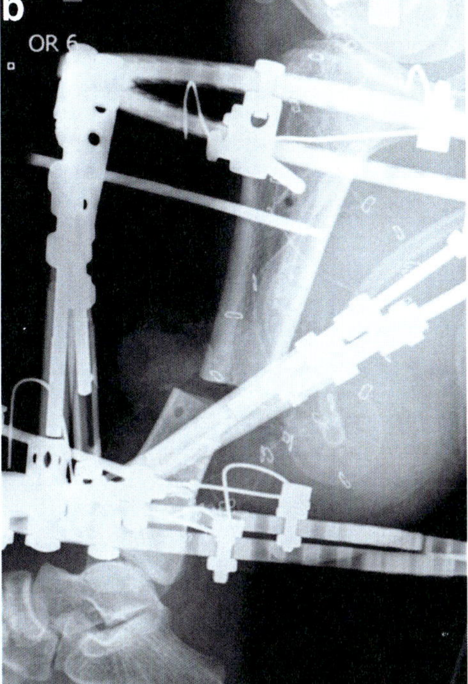


**Fig. 3** Modern ring fixator to treat high-energy open fracture (**a**) Photograph of ring fixator. (**b**) Radiographic image of ring fixator. A "soft tissue reduction" was used to shorten the limb, obtain adequate soft tissue closure, and then lengthen the limb out slowly over time to restore length and alignment

however, these infections can typically be treated with oral antibiotics and are much less severe than osteomyelitis that develops at the fracture site.

## *3.1 Incident Infection with Internal vs. External Fixation for Severe Tibial Fractures*

A review of 29 studies evaluating the treatment of open tibial fractures reported a lower infection rate when using external fixation compared with internal fixation

**Table 4** Combat-related tibial fractures

| Reference | Design | Subjects | Infection outcome |
|---|---|---|---|
| Internal fixation | | | |
| Lacap and Frisch (2007) | Observational study | 35 Type III wartime tibial fractures treated with intramedullary nail fixation | Deep infections in 14.3% with Type III fractures |
| External fixation | | | |
| Zeliko et al. (2006) | Observational, 6-month study from 1991 to 1992 | 49 wounded patients with war injuries to extremities including open fractures, 8 required amputation, 27 were treated with external fixation, and 8 with internal fixation | Osteomyelitis occurred in five patients, only one of whom was treated with primary external fixation |
| Lerner et al. (2006) | Retrospective review | 64 high-energy limb fractures caused by war injuries in 47 patients treated with staged external fixation | Osteomyelitis or deep infection in 6.3% |
| Keeling et al. (2008) | Retrospective review | 67 Type III tibia fractures obtained during combat in 65 patients treated with ring external fixation | Deep infections in 7.9% overall and 2.9% with Type IIIC fractures |

using plates and screws (Giannoudis et al. 2006). Infections were least frequent with tibial nailing. Direct comparisons across treatments, however, are limited as these data were gathered from a range of studies that included potentially important differences in rates of pin tract infections (which are of little consequence) versus osteomyelitis, patient populations, wound/injury type, level of contamination, and other complications. In particular, it may be especially important to distinguish outcome in tibial fractures obtained through civilian or combat-related injuries, which often involve higher velocity injuries with more contamination.

### 3.1.1 Infections with Tibia Fractures with Higher Energy Combat and Civilian Injuries

Military open tibial fracture is currently often treated definitively with external fixation at US centers, although data supporting this practice are somewhat limited. Several studies have reported infection incidence in patients treated for severe, combat-related tibial fractures (Lacap and Frisch 2007; Zeliko et al. 2006; Lerner et al. 2006; Keeling et al. 2008; Table 4). In general, these data support a lower risk of deep infection and osteomyelitis when combat patients are treated with external fixation.

Low infection rates have also been reported with noncombat tibial fractures treated with ring external fixation in very high-energy trauma patients. A retrospective review of a civilian sample evaluated outcome with external fixation over a 16-year period at a single trauma center (Hutson et al. 2010). Seventy-six patients with 78 Type IIIB tibial fractures were included. Eighteen patients (19 fractures) had bone defects requiring external fixation with distraction osteogenesis. There were no cases of deep infection reported in this sample, which is remarkable given the high-energy nature of the cohort. This may be due to the lack of metal at the fracture site for bacteria to form biofilm, better debridement, or other factors that could be specific to this single site. Further study is needed to see if these results can be generalized.

### 3.1.2 Randomized Data Comparing Internal vs. External Fixation

Most studies reporting infection outcome in patients treated with internal or external fixation are based on observational data in which treating surgeons selected fixation method or all patients were treated with the same treatment. Limited data are available directly comparing outcome with internal vs. external fixation. A prospective trial randomized 59 patients with Types II or III tibial shaft fractures from civilian injuries (most commonly motor traffic accidents) to tibial shaft plating or external fixation (Bach and Hansen 1989). Seventeen fractures were Type III. Wound infection and osteomyelitis were more common among patients treated with internal fixation. This study is typical of the few randomized controlled series on this topic as it is underpowered to address the primary outcome measure of infection.

## 4 Prospective FixIt Trial

The FixIt trial is a prospective, randomized trial designed to assess fixation strategies for severe open tibial fractures using internal fixation (with nails or plates) or external ring fixation (Personal communication O'Toole. Study description available at http://metrc.org/; accessed July 2011). The FixIt trial is planned to be conducted in 23 major trauma centers in the United States, including four military treatment centers. The goal is to randomize up to 300 patients with severe open tibial shaft fractures and follow these patients for 1 year. Inclusion criteria include Type IIIB or "severe" IIIA tibial fractures. IIIA fractures will be considered to be severe if they are too contaminated for intramedullary nails fixation on the first trip to the operating room, require split-thickness skin graft, would meet other criteria for IIIB except that the wound is closed, or include a bone gap >1 cm. The primary outcome measure is re-admission for complication, with incident infection an important secondary outcome measure. It is hoped that data from this study will help provide additional information linking infection risk to fixation method in both civilian and combat tibial fractures.

## *4.1 Challenges for the FixIt Trial*

Randomized trials are needed to provide the most reliable comparative data between treatments to avoid selection biases in treatment assignment that might be expected to also influence infection risk. While randomization provides a better study design, surgeons and patients involved in the study must agree to participate in the randomization process. It is currently unknown if it will be possible to enroll an adequate number of patients and surgeons for such a trial.

Collecting an adequate sample will likely take substantial time, as severe tibial fractures are not common. For example, over a 5-year period, only 74 tibial fracture patients requiring flap coverage were treated at the R. Adams Cowley Shock Trauma Center at the University of Maryland School of Medicine, one of the largest US trauma centers (D'Alleyrand et al. 2010). Lack of a sufficient sample size at a single center makes a multicenter design attractive for the FixIt trial.

# 5 Conclusion

Additional data are needed to make confident recommendations for the fixation strategy of severe open tibial fractures to reduce infection risk. Although the currently available literature might be interpreted to support reduced infection with external fixation, interpretations that can be made from observational and retrospective data are limited by selection bias that likely affects outcome. The large, prospective FixIt trial may provide important information based on its randomized design and planned substantial sample size obtained from multiple centers.

# References

Bach AW, Hansen ST (1989) Plate versus external fixation in severe open tibial shaft fractures. Clin Orthop Relat Res 241:89–94

Bosse MJ, MacKenzie EJ, Kellam JF et al (2002) An analysis of outcomes of reconstruction or amputation of leg-threatening injuries. N Engl J Med 347:1924–1931

Burns TC, Stinner DJ, Possley DR et al (2010) Does the zone of injury in combat-related Type III open tibia fractures preclude the use of local soft tissue coverage? J Orthop Trauma 24:697–703

D'Alleyrand JG, Dancy L, Castillo R, et al (2010) Is time to flap coverage an independent predictor of flap complication. In: Presented at the Orthopaedic Trauma Association 26th annual meeting, October 16, Baltimore, MD. http://www.hwbf.org/ota/am/ota10/otapa/OTA100671.htm. Accessed July 2011

Doucet JJ, Galameau MR, Potenza BM et al (2011) Combat versus civilian open tibia fractures: the effect of blast mechanism on limb salvage. J Trauma 70:1241–1247

Giannoudis PV, Papkostidis C, Roberts C (2006) A review of the management of open fractures of the tibia and femur. J Bone Joint Surg Br 88:281–289

Gottenbos B, van der Mei HC, Busscher HJ (2000) Initial adhesion and surface growth of *Staphylococcus epidermidis* and *Pseudomonas aeruginosa* on biomedical polymers. J Biomed Mater Res 50:208–214

Gustilo RB, Mendoza RM, Williams DN (1984) Problems in the management of Type III (severe) open fractures: a new classification of Type III open fractures. J Trauma 24:742–746

Hutson JJ, Dayicioglu D, Oeltjen JC et al (2010) The treatment of Gustilo grade IIIB tibia fractures with application of antibiotic spacer, flap, and sequential distraction osteogenesis. Ann Plast Surg 64:541–552

Keeling JJ, Gwinn DE, Tintle SM et al (2008) Short-term outcomes of severe open wartime tibial fractures treated with ring external fixation. J Bone Joint Surg Am 90:2643–2651

Lacap AP, Frisch HM (2007) Intramedullary nailing following external fixation in tibial shaft fractures sustained in Operations Enduring and Iraqi Freedom. In: Poster presented at the Annual Meeting of the Orthopaedic Trauma Association, Oct 17–20, Boston, MA

Lerner A, Fodor L, Soudry M (2006) Is staged external fixation a valuable strategy for war injuries to the limbs? Clin Orthop Relat Res 448:217–224

Stall A, Gupta R, O'Toole RV, Zadnik M (2010) Does perioperative hyperoxygenation decrease surgical site infection in orthopaedic trauma patients? Podium presentation at American Academy of Orthopaedic Surgeons, March, New Orleans, LA

van der Borden AJ, van der Werf H, van der Mei HC, Busscher HJ (2004) Electric current-induced detachment of *Staphylococcus epidermidis* biofilms from surgical stainless steel. Appl Environ Microbiol 70:6871–6874

Zeliko B, Lovrć Z, Amć E et al (2006) War injuries of the extremities: twelve-year follow-up data. Mil Med 171:55–57

# Second-Generation Molecular Diagnostics and Strategies for Preventing Periprosthetic Joint Infections

Nicholas Sotereanos

**Abstract** Bacteria are highly successful organisms, with biofilms enhancing their ability to effectively mutate to acquire drug resistance (Harbottle et al. 2006; Høiby et al. 2010; Jolivet-Gougeon et al. 2011). Fossil records show that bacteria have been using biofilm survival mechanisms for over three billion years (Hall-Stoodley et al. 2004; Tice and Lower 2004). The extensive history of bacteria in perfecting biofilms has resulted in a long record of success for bacteria in natural ecosystems, and to problems for clinicians in eradicating biofilm infections.

Orthopedists need to understand that bacterial biofilms have evolved to become successful survivors, displaying aggressive and resistant qualities. Biofilms are often difficult to detect using traditional diagnostic techniques. In addition, the inherent resistance of biofilms to antimicrobials has made chronic bacterial infections a major problem facing orthopedic surgeons. Periprosthetic joint infections (PJIs) occur in 1–2% of primary total knee and hip arthroplasties and 3–9% of revisions (Blom et al. 2004; Ridgeway et al. 2005; Wilson et al. 2008; Mortazavi et al. 2010). Although only a minority of arthroplasties are complicated by PJI, current treatment of established PJI biofilms in orthopedics requires extensive and often debilitating surgeries. Consequently, orthopedic infection management should focus on early recognition of infection and on strategies to reduce infection risk before patients come into the operating room.

## 1 Limitations of Traditional Testing for Orthopedic Infections

Clinicians have been trained to rely on culture results to guide the diagnosis and treatment of infection. In 1884, Dr. Koch published postulates for identifying pathogens for infection (Koch 1884; Box 1). These criteria founded diagnostic

N. Sotereanos (✉)
Department of Orthopedic Surgery, Allegheny General Hospital, Pittsburgh, PA, USA
e-mail: nsotereanos@usa.net

G.D. Ehrlich et al. (eds.), *Culture Negative Orthopedic Biofilm Infections*,
Springer Series on Biofilms 7, DOI 10.1007/978-3-642-29554-6_8,

principles that formed the foundation of early microbiology and continue to drive clinical diagnoses today (Koch 1884; Carter 1985; Inglis 2007).

Koch's postulates rely on isolating and growing bacteria in culture. While clinicians are most comfortable when culture results are able to drive infection treatment, PJIs often fail to produce positive cultures in patients with clinical infections (see chapter "Diagnosing Periprosthetic Joint Infection: Cultures, Molecular Markers, and the Ibis Technology" by Parvizi). For example, a retrospective survey of 897 patients with PJIs reported that 7% occurred in patients whose initial cultures were negative (Berbari et al. 2007). Koch also recognized the limitations of his postulates, including the issue of culture-positive individuals without clinical infection who are merely colonized by organisms rather than infected. The development of molecular techniques that may identify microbial pathogens missed with routine culture has called into question the need to fulfill all of Koch's postulates before diagnosing infections (Post et al. 1995; Fredricks and Relman 1996).

## 1.1 Diagnostic Obstacles with Currently Used Diagnostic Tools

The diagnosis of PJI generally relies on the presence of clinical features (e.g., rest pain) and laboratory evidence of inflammation (e.g., elevated white blood cell counts, C-reactive protein, and erythrocyte sedimentation rate). Traditional cultures are frequently negative in patients with clinical evidence for PJI, with results often confounded by preoperative antibiotic treatment. Additional tools for diagnosing infection have also proved unreliable for PJIs. Intraoperative Gram-stain results, relying on the presence of organisms or high counts of neutrophils, also frequently miss PJIs (Spangehl et al. 1999; Morgan et al. 2009; Ghanem et al. 2009; Table 1). For example, the test accuracy of intraoperative Gram-stains for patients undergoing revision total knee arthroplasty is 80% (Morgan et al. 2009). In addition, nuclear medicine testing frequently fails to identify PJI (Love et al. 2009). Indium-111 white blood cell scintigraphy has been shown to be sensitive for identifying PJI after total hip and knee replacements; however, specificity is only 80% for hip and 50% for knee replacement PJIs (Nijhof et al. 1997).

**Table 1** Predictive value of intraoperative Gram stains for PJI

| | Total hip arthroplasty | | Total knee arthroplasty | |
|---|---|---|---|---|
| | $N = 202$ (Spangehl et al. 1999) | $N = 651$ (Ghanem et al. 2009) | $N = 921$ (Morgan et al. 2009) | $N = 453$ (Ghanem et al. 2009) |
| Sensitivity (%) | 19 | 43 | 27 | 64 |
| Specificity (%) | 98 | 99 | 100 | 99 |
| Positive predictive value (%) | 63 | 93 | 99 | 97 |
| Negative predictive value (%) | 89 | 82 | 79 | 82 |

**Box 1. Koch's Postulates**

1. Infected tissue must show the presence of a particular microorganism not found in a healthy organism.
2. The microorganism must be isolated and grown in a pure culture.
3. When injected into a healthy organism, the microorganism must cause the disease associated with it.
4. This microorganism should then be able to be isolated and shown to be identical to the microorganism found in an infected individual.

## 2 Molecular Diagnostics

First-generation molecular diagnostics are based on polymerase chain reaction (PCR) technology. Despite success in identifying infections in some culture-negative patients, PCR techniques are limited by false negatives from testing flaws and false positives from contaminants (Ehrlich 2011). Second-generation molecular tests have improved on earlier diagnostics, including more extensive identification of genetic sequences against which samples may be compared to find bacterial matches and the addition of mass spectroscopy and other techniques to PCR technology to improve both sensitivity and specificity for PJI diagnosis.

### *2.1 Defining Common Techniques*

Second-generation molecular techniques for detecting and identifying bacterial infections use extracted and purified genetic sequences (Costerton et al. 2011). Important sequences for bacterial species identification include the 16S ribosomal gene that codes the 16S rRNA that provides a bacterial "fingerprint" and various conserved and taxon-specific diagnostic and virulence genes. The second-generation techniques can also detect antibiotic resistance genes like *mecA* for methicillin resistance and *van* A–C that code for vancomycin resistance. Gene sequences are amplified using PCR, which copies segments of genetic material. Amplified gene sequences (called amplicons) can then be evaluated using electrophoresis or mass spectroscopy. Bacterial infections, therefore, might be uncovered by probing for commonly encountered pathogens in specific patient populations.

Fluorescence in situ hybridization (FISH) is a specific second-generation molecular technique that has been successfully applied to orthopedic biofilm infections. FISH originally used fluorescent DNA probes that attach to genetic material of the human chromosome (Fig. 1), and subsequently was adapted to study bacterial 16S rRNA. Biofilm structure can then be more carefully explored by FISH through three-dimensional confocal microscopic imaging (Fig. 2). These molecular techniques have been used to show the presence of cells of both *Staphylococcus*

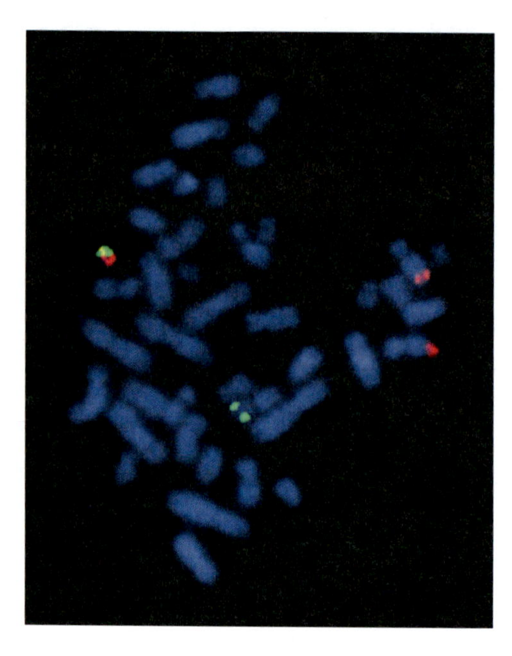

**Fig. 1** FISH can identify components of human chromosomes when single-stranded DNA sequences are labeled with fluorescent chromophores (*red* and *green*) and allowed to "find" corresponding genetic sequences in the chromosomes of a cell undergoing mitosis

*aureus* and *Staphylococcus epidermidis* in perioperative tissues obtained from culture-negative joint infections (Stoodley et al. 2005).

The Ibis technology is another second-generation molecular technology in which PCR-derived amplicons are evaluated through mass spectroscopy to help identify both commonly encountered pathogens and other agents that might not have been anticipated. For example, identifying an amplicon that fails to match a pathogen available in the database genomic library provides important information that additional agents are present. Ibis is able to detect bacteria, as well as fungi and viruses. The Ibis technology can more rapidly identify pathogens than traditional culture, with results expected in <6 h.

## 2.2 Orthopedic Case Using Standard Culture and Molecular Techniques

The benefits of molecular diagnostics over standard culture methods are highlighted by the published case of a 34-year-old man treated for an open tibial fracture, complicated by delayed healing and periodic pain (Palmer et al. 2011). The patient was originally treated with debridement and external fixation, followed by intramedullary nailing. Six months later, delayed healing resulted in dynamization of the nail. Twenty-two months after surgery, increasing pain resulted in obtaining

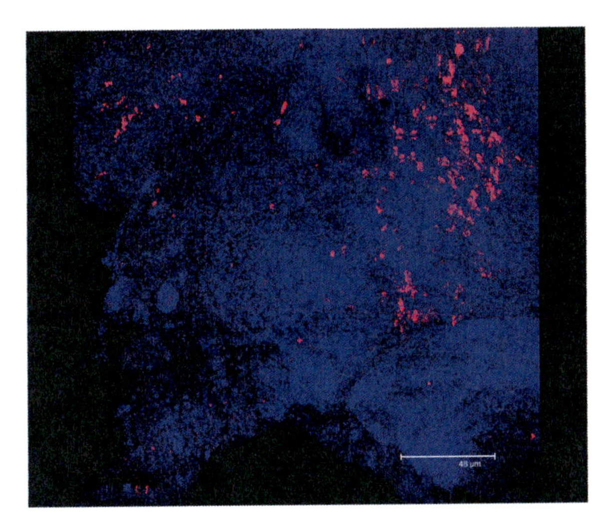

**Fig. 2** FISH be used to identify bacteria *in situ* in human tissues. Single-stranded rDNA sequences are labeled with fluorophores and allowed to hybridize to complementary regions of the 16S rRNA of bacterial ribosomes that are characteristic of particular species. Here cells of a Staphylococcal species are visualized in the tissues adjacent to a nonunion that was culture negative

radiographs that continued to show limited fracture healing. Serum inflammatory markers were not elevated and cultures were negative. Ibis testing identified staphylococcal bacteria and the *mecA* gene for methicillin resistance. Bacterial presence was further confirmed through FISH testing of tissue samples Fig. 2. This case highlights the ability of second-generation molecular testing to uncover clinically important infections not identified through routine clinical evaluations.

## 2.3 Limitations of Techniques and Challenges for the Surgeon

While second-generation molecular techniques offer promise for improving the diagnosis of PJI, the practical usefulness of these tests is limited. Surgeons will need tests that are both sensitive and specific that can provide immediate or ultra-rapid diagnoses in the operating room. Even tests, like Ibis, that offer more rapid diagnostic results (<6 h) than traditional cultures fail to provide results quickly enough to aid in decision making in the operating room. Clinically practical tests will need to produce results within minutes rather than hours to help guide surgery and direct surgeons to areas from which they might obtain biopsies that will be most likely to produce useful pathogen identification through culture.

# 3 Implementing Programs for Proactive Infection Control

Orthopedic management of established PJIs may require additional surgery, implant removal, and potentially significant postoperative morbidity. Proactive techniques designed to identify and reverse modifiable risk factors for PJI are essential strategies for infection control to avoid costly and potentially disabling revisions. Reducing infection risk by recognizing the frequent asymptomatic colonization with pathogens that may pose surgical risk, pretreating patients with antiseptic washes, and reducing the introduction of pathogens into the operating room (by limiting the number of individuals present and the movement of potentially contaminated air into the operating room) are all strategies to help reduce PJI incidence.

## *3.1 Impact of Pathogen Risk from Operating Room Staff*

Operating room staff can become important vectors of infection that increase the risk for patients acquiring surgical wound infections (Ayliffe 1991). Operating staff need to be educated about their role as carriers of pathogens that might contaminate the surgical environment. For example, bacterial counts increase dramatically once staff are added to an operating room. In an interesting study, the number of airborne bacteria collected in a horizontal-flow, bioclean operating room increased 26 times from 0.1 colony-forming unit per cubic meter in an empty waiting room to 2.6 colony-forming units per cubic meter when an average of 8 staff members were present in the room (Suzuki et al. 1984). Bacteria were predominantly Gram-positive cocci, most commonly coagulase-negative staphylococci (61% of bacteria collected). Strategies to reduce contamination from staff are an essential component of proactive infection control.

### 3.1.1 Impact of Staff Head Covering

While head coverings are commonly used in the operating room, some styles of head covering produce complete hair covering, while others only achieve partial covering (Fig. 3). In a study evaluating the effect of different head coverings on surgical area contamination, the type of head covering was less important than the use of a head cover that provided complete head covering (Friberg et al. 2001). Staff showered and washed their hair prior to participation and all participants wore standard operating room gowns, sterile surgical gloves, and a mask. Results were similar when using either a nonsterile squire-type hood that reached under the gown or sterilized helmet aspirator system; however, bacterial contamination of surfaces increased three- to fivefold when no head covering was used and wound area contamination increased 60-fold. Most organisms were Streptococci, presumed to be of respiratory origin.

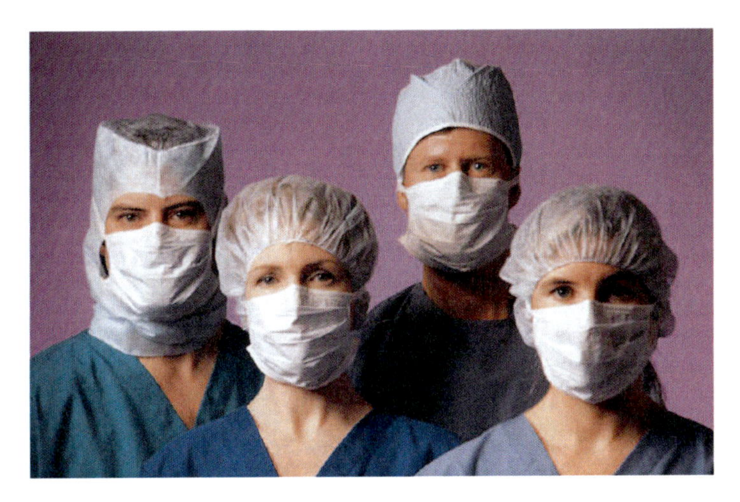

**Fig. 3** Surgical cap styles. The cap on the *left* provides more complete hair covering than traditional or bouffant caps

## 3.2 Colonization with Important Pathogens

Colonization with pathogens is an important factor increasing risk for the development of clinical infection. A systematic review recently determined that risk of infection was increased fourfold among individuals colonized with methicillin-resistant *Staphylococcus aureus* (MRSA) (Safdar and Bradley 2008). Colonization of both staff and patients may be important for risk of PJIs in the operating room. Unfortunately, most operating room staff and patients are unaware that they may be colonized and that colonization is a significant risk factor.

### 3.2.1 Colonization Among Patients

Asymptomatic nasal colonization with MRSA has been reported not uncommonly. Using laboratory cultures from nasal swabs obtained from 758 patients at hospital admission, 3% were colonized with MRSA and 21% with methicillin-sensitive *S. aureus* (Davis et al. 2004). Incident infection during hospitalization was more common among patients who had been colonized with MRSA at admission (Fig. 4). In another study, PCR was used to identify *S. aureus* from samples taken from the anterior nares, oropharynx, palms, groin, perirectal area, wounds, and catheter insertion sites of 400 patients presenting to an emergency department (Schechter-Perkins et al. 2011). Methicillin-sensitive *S. aureus* was present in 39% of patients, with 5% colonized with MRSA.

The significance of identifying preoperative colonization for PJI was highlighted in a study implementing colonization identification and decolonization prior to orthopedic procedures. Asymptomatic nasal colonization with MRSA was evaluated through a surveillance prescreening program using PCR technology in

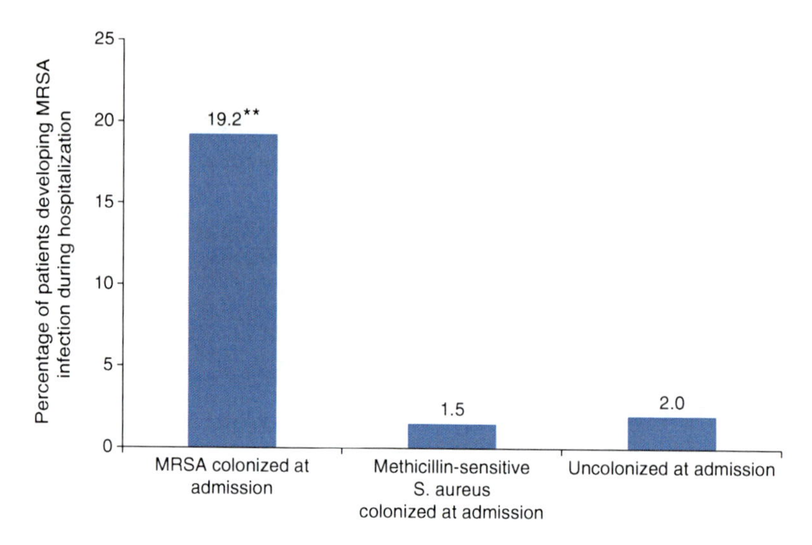


**Fig. 4** Risk for developing MRSA during hospitalization was significantly higher among patients with asymptomatic nasal colonization with MRSA at hospital admission. Difference between samples was significant for MRSA vs. methicillin-sensitive and uncolonized groups, $^{**}P < 0.01$ (based on Davis et al. 2004)

patients scheduled for orthopedic surgery (Spencer et al. 2008). MRSA-positive screened patients were treated with 5-day intranasal 2% mupirocin twice daily and daily cleansing with chlorhexidine and re-screen before surgery. These patients were also treated with vancomycin for surgical prophylaxis. Among the 7,019 patients screened, *S. aureus* was identified in 23% of patients, and 4% were MRSA-positive. Repeated screens were negative for MRSA in 78% who had previously been positive. Prior to implementation of this screening and decolonization program, surgical infections were reported in 0.46% of orthopedic inpatients. Following the screening and decolonization program, infection rate was 0.18% overall, with most infections occurring in those patients identified before surgery with nasal MRSA (Fig. 5). Therefore, preoperative identification of *S. aureus* and MRSA linked with decolonization procedures successfully reduced infection.

### 3.2.2 Colonization Among Hospital Staff

Operating room staff may also be important vectors of infection. A survey screening 100 surgical unit workers (42 doctors, 30 residents, 25 nurses, 40 medical students, and 30 nursing students) with nasal swabs indentified *S. aureus* colonization in 13% and coagulase-negative *Staphylococcus* in 63% (Fig. 6; Vinodhkumatadithyaa et al. 2009). MRSA was identified in 2 of the 13 staff positive for *S. aureus*. A similar screening for nasal colonization reported MRSA in 15% of 105 emergency department staff members (18% of technicians were positive for nasal colonization with MRSA, 17% of nurses, and 8% of doctors; Bisaga et al. 2008).

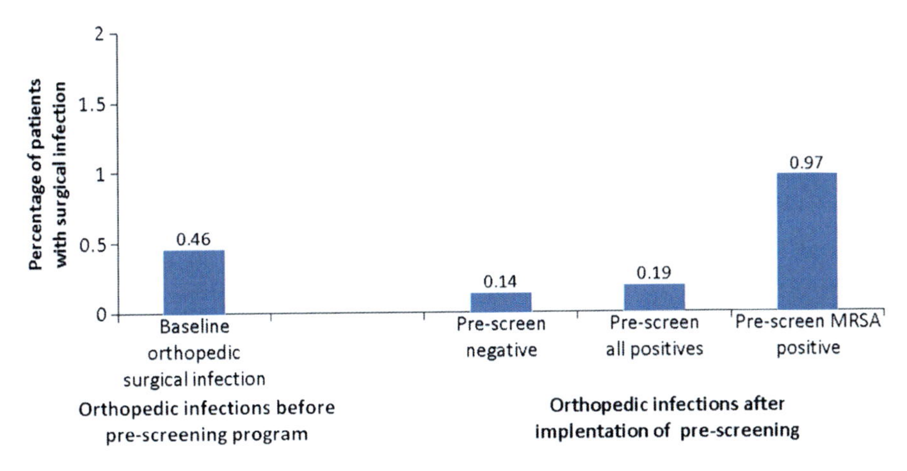


**Fig. 5** Orthopedic surgery infection incidence before and after implementing a preoperative screening and decolonization program (based on Spencer et al. 2008)

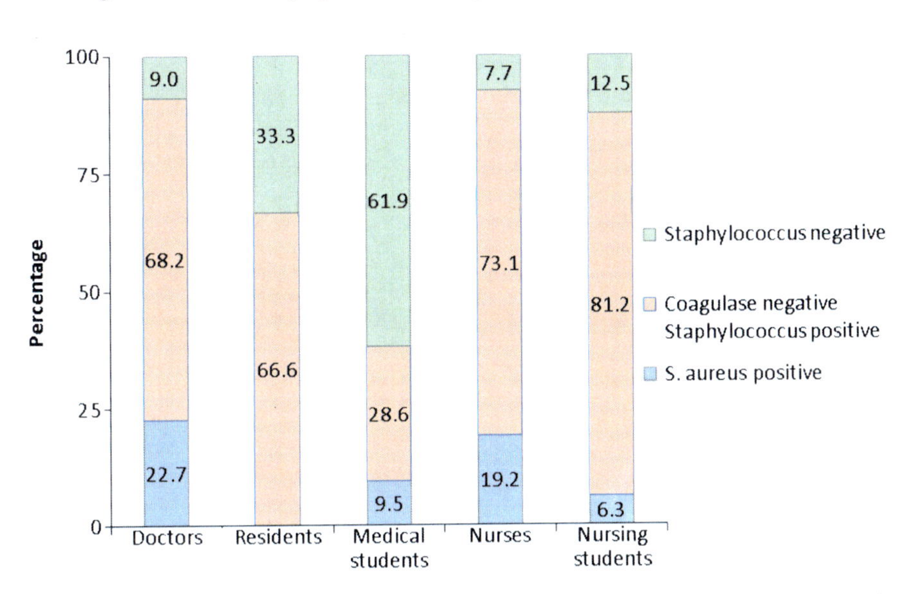


**Fig. 6** Colonization with staphylococci in hospital staff (based on Vinodhkumatadithyaa et al. 2009)

## *3.3 Preoperative Chlorhexidine*

Chlorhexidine is an antiseptic skin wash. In a retrospective study, 954 patients scheduled for hip arthroplasty were evaluated to compare infection rates between those completing at-home, preoperative chlorhexidine washes the night before and the morning of surgery and those who did not use preoperative chlorhexidine and received only standard perioperative skin cleansing (Johnson et al. 2010). Patient

**Table 2** Strategies to reduce infection risk in the operating room

| Preoperative strategies | Staff strategies | Operating room procedures |
|---|---|---|
| Prescreening for pathogen colonization, followed by decolonization strategies | Require staff to bath prior to each day in the operating room | Maintain operating room ventilation system |
| Chlorhexidine showers 3 days prior to and the day of scheduled surgery | Require complete hair coverings at all times in the operating room | Limit the number of personnel in the operating room to essential staff |
| | Rescrub if leaving and then re-entering the operating room | Avoid opening doors during surgeries |
| | | Concentrate airflow over the operative site rather than the entire operating room |
| | | Patient skin should be covered except for the area necessary to be uncovered for procedures |

outcome was stratified based on infection risk categories using the National Nosocomial Infections Surveillance System surgical risk classification that considers both surgical incision time and wound contamination. There were no infections among any of the patients using chlorhexidine; infection occurred in 2% of low-risk, 3% of the medium-risk, and 7% of the high-risk patients using only perioperative cleansing.

Another study evaluated infection rates for patients using 2% chlorhexidine-impregnated cloths the night before and the morning of their scheduled elective knee arthroplasty to disinfect their neck, chest, abdomen, back, and extremities compared with patients using in-hospital preparation only (Zywiel et al. 2011). There were no surgical site infections among the 136 patients using the preoperative chlorhexidine washes vs. 21 infections in the 711 patients having surgery using only perioperative cleansing (3%).

## 3.4 *Developing Practical Strategies for Reducing Surgical Infections*

Simple strategies supported by the data above generally add minimal additional cost to running an operating room and are recommended to help reduce infection risk (Table 2). A quality improvement program implementing these strategies through multidisciplinary teams that involved surgical staff, infectious disease physicians, and pharmacists in 56 participating centers resulted in a 27% reduction in surgical site infections (Colorado Foundation for Medical Care website 2011). Operating room staff should be educated to understand that they can become important vectors bringing pathogens into the operating room. Limiting the number of personnel and restricting traveling into and out of the operating room during procedures can reduce pathogen burden. In addition, meticulous hygiene and using clean head coverings that provide total hair coverage are important strategies to reduce operating room contamination from staff. Voluntary screening of nasal passages for MRSA and perirectal swabbing for vancomycin-resistant enterococci every 6 months for operative staff should also be considered. Identifying commonly encountered pathogens and cataloguing their genomes may help guide future patient care.

Although data are limited, laminar air flow and ultraviolet light have both been shown in small samples to reduce PJIs (Evans 2011). While laminar air flow warrants additional studies to confirm benefits, the Centers for Disease Control recommend against using ultraviolet light, due to potential health risks for exposed healthcare personnel (Evans 2011). Data collected in over 51,000 total hip and 36,000 total knee replacements failed to show reduction in early deep infections from using either laminar flow or space suits (Hooper et al. 2011).

### 3.4.1 Antibiotic Treatment

Antibiotic coverage can usually be based on results from previous cultures of organisms commonly encountered in different environments. Coverage may need to be broader among patients with conditions that predispose them to getting PJI (Box 2). For example, patients with rheumatoid arthritis are at higher risk for PJI than those with osteoarthritis. Using a Mayo Clinic sample of patients with arthritis undergoing total hip or knee arthroplasties, PJIs occurred in 1% of patients with osteoarthritis and 4% with rheumatoid arthritis (Bongartz et al. 2008). Among the rheumatoid arthritis patients, PJI occurred in 2% of primary surgeries and 6% of revisions. In addition, patients who have spent several days in the hospital before surgery may be at higher risk because their time in the hospital may have resulted in colonization with additional pathogens, including resistant bacteria.

**Box 2. Conditions Suggesting Patients Who Are at Higher Risk for Developing Surgical Infection**

- Obesity
- Psoriasis
- Diabetes
- Rheumatoid arthritis
- Steroid use
- Urinary tract infection
- Malnutrition (e.g., serum albumin <3.5 g/dL)
- Hospitalization prior to surgery

# 4 Conclusions

Pre- and intraoperative diagnosis of PJI is complicated by limitations in sensitivity from traditional methods for identifying infection. Cultures, laboratory testing, and nuclear medicine scans often fail to identify clinically important PJI. While second-generation molecular diagnostics provide tests with greater sensitivity, delays in receiving results limits their usefulness in typical clinical situations.

The difficulty in diagnosing and eradicating these biofilm infections makes proactive prevention strategies of PJI especially important. Proactive PJI control focuses on understanding the important role that colonization of both staff and patients can have on introducing pathogens into the operating room. Strategies to identify and decolonize pathogens before surgery may help reduce PJIs and the potentially disabling surgeries needed to eradicate PJI.

# References

Ayliffe GA (1991) Role of the environment of the operating suite in surgical wound infection. Rev Infect Dis 13(Suppl 10):S800–S804

Berbari EF, Marculescu C, Sia I et al (2007) Culture-negative prosthetic joint infection. Clin Infect Dis 45:1113–1119

Bisaga A, Paquette K, Sabatini L, Lovell EO (2008) A prevalence study of methicillin-resistant *Staphylococcus aureus* colonization in emergency department health care workers. Ann Emerg Med 52:525–528

Blom AW, Brown J, Taylor AH et al (2004) Infection after total knee arthroplasty. J Bone Join Surg Br 86:688–691

Bongartz T, Halligan CS, Osmon DR et al (2008) Incidence and risk factors of prosthetic joint infection after total hip or knee replacement in patients with rheumatoid arthritis. Arthritis Rheum 59:1713–1720

Carter KC (1985) Koch's postulates in relation to the work of Jacob Henle and Edwin Klebs. Med Hist 29:353–374

Colorado Foundation for Medical Care website. Surgical care improvement project. http://www.cfmc.org/hospital/hospital_scip.htm. Accessed Aug 2011

Costerton JW, Post JC, Ehrlich GD et al (2011) New methods for the detection of orthopedic and other biofilm infections. FEMS Immunol Med Microbiol 61:133–140

Davis KA, Stewart JJ, Crouch HK, Florez CE, Hospenthal DR (2004) Methicillin-resistant *Staphylococcus aureus* (MRSA) nares colonization at hospital admission and its effect on subsequent MRSA infection. Clin Infect Dis 39(6):776–782

Ehrlich GD (2011) Next generation molecular diagnostics for orthopedic infections. J Bone Joint Surg (Br) 93B(Suppl III):318

Evans RP (2011) Current concepts for clean air and total joint arthroplasty: laminar airflow and ultraviolet radiation: a systematic review. Clin Orthop Relat Res 469:945–953

Fredricks DN, Relman DA (1996) Sequence-based identification of microbial pathogens: a reconsideration of Koch's postulates. Clin Microbiol Rev 9:18–33

Friberg B, Friberg S, Östensson R, Burman LG (2001) Surgical area contamination–comparable bacterial counts using disposable head and mask and helmet aspirator system, but dramatic increase upon omission of head-gear: an experiment study in horizontal laminar air-flow. J Hosp Infect 47:110–115

Ghanem E, Ketonis C, Restrepo C et al (2009) Periprosthetic infection: where do we stand with regard to Gram Stain? Acta Orthop 80:37–40

Hall-Stoodley L, Costerton JW, Stoodley P (2004) Bacterial biofilms: from the environment to infectious disease. Nat Rev Microbiol 2:95–108

Harbottle H, Thakur S, Zhao S, White DG (2006) Genetics of antimicrobial resistance. Anim Biotechnol 17:111–124

Høiby N, Bjarnsholt T, Givskov M, Molin S, Ciofu O (2010) Antibiotic resistance of bacterial biofilms. Int J Antimicrob Agents 35:322–332

Hooper GJ, Rothwell AG, Frampton C, Wyatt MC (2011) Does the use of laminar flow and space suits reduce early deep infection after total hip and knee replacement?: the ten-year results of the New Zealand Joint Registry. J Bone Joint Surg Br 93:85–90

Inglis TJ (2007) Principia aetiologica: taking causality beyond Koch's postulates. J Med Microbiol 56:1419–1422

Johnson AJ, Daley JA, Zywiel MG, Delanois RE, Mont MA (2010) Preoperative chlorhexidine preparation and the incidence of surgical site infections after hip arthroplasty. J Arthroplasty 26 (suppl 6):98–102

Jolivet-Gougeon A, Kovacs B, Le Gall-David S et al (2011) Bacterial hypermutation: clinical implications. J Med Microbiol 60:563–573

Koch R (1884) Die Aetiologie der Tuberkulose. Mitt Kaiser Gesundh 2:1–88

Love C, Marwin SE, Palestro CJ (2009) Nuclear medicine and the infected joint replacement. Semin Nucl Med 39:66–78

Morgan PM, Sharkey P, Ghanem E et al (2009) The value of intraoperative Gram stain in revision total knee arthroplasty. J Bone Joint Surg Am 91:2124–2129

Mortazavi SM, Schwartzenberger J, Austin MS, Purtill JJ, Parvisi J (2010) Revision total knee arthroplasty infection: incidence and predictors. Clin Orthop Relat Res 468:2052–2059

Nijhof MW, Oyen WJ, van Kampen A et al (1997) Hip and knee arthroplasty infection. In-111-IgG scintigraphy in 102 cases. Acta Orthop Scan 68:332–336

Palmer M, Costerton W, Sewecke J, Altman D (2011) Molecular techniques to detect biofilm bacteria in long bone nonunion: a case report. Clin Orthop Relat Res 469:3037–3042

Post JC, Preston RA, Aul JJ, Larkins-Pettigrew M, Rydquist-White J, Anderson KW, Wadowsky RM, Reagan DR, Walker ES, Kingsley LA, Magit AE, Ehrlich GD (1995) Molecular analysis of bacterial pathogens in otitis media with effusion. JAMA 273: 1598–1604.

Ridgeway S, Wilson J, Charlet A et al (2005) Infection of the surgical site after arthroplasty of the hip. J Bone Joint Surg Br 87:844–850

Safdar N, Bradley EA (2008) The risk of infection after nasal colonization with *Staphylococcus aureus*. Am J Med 121:310–315

Schechter-Perkins EM, Mitchell PM, Murray KA et al (2011) Prevalence and predictors of nasal and extranasal staphylococcal colonization in patients presenting to the emergency department. Ann Emerg Med 57:492–499

Spangehl MJ, Masterson E, Masri BA, O'Connell JX, Duncan CP (1999) The role of intra-operative Gran stain in the diagnosis of infection during revision total hip arthroplasty. J Arthroplasty 14:952–956

Spencer MP, Gulcynski D, Davidson S, Richmond J (2008) Eradication of methicillin sensitive *Staphylococcus aureus* and methicillin resistant *Staphylococcus aureus* before orthopedic surgery. In: Presented at the 18th Annual Scientific Meeting of the Society for Healthcare Epidemiology of America (SHEA), April 5-8, 2008, Orlando, FL

Stoodley P, Kathju S, Hu FZ et al (2005) Molecular and imaging techniques for bacterial biofilms in joint arthroplasty infections. Clin Orthop Relat Res 437:31–40

Suzuki A, Namba Y, Matsuura M, Horisawa A (1984) Airborne contamination in an operating suite: report of a five-year survey. J Hyg Camb 93:567–573

Tice MM, Lower DR (2004) Photosynthetic microbial mats in the 3,416-Myr-old ocean. Nature 431:549–552

Vinodhkumatadithyaa A, Uma A, Srinivasan M et al (2009) Nasal carriage of methicillin-resistant *Staphylococcus aureus* among surgical unit staff. Jpn J Infect Dis 62:228–229
Wilson J, Charlett A, Leong G, McDougall C, Duckworth G (2008) Rates of surgical site infection after hip replacement as a hospital performance indicator: analysis of data from the English mandatory surveillance system. Infect Control Hosp Epidemiol 29:219–226
Zywiel MG, Daley JA, Delanois RE et al (2011) Advance pre-operative chlorhexidine reduces the incidence of surgical site infections in knee arthroplasty. Int Orthop 35:1001–1006

# Treatment of Orthopedic Infections: Addressing the Biofilm Issue

**Heinz Winkler**

**Abstract** Chronic infections of bone and joints always have been considered especially difficult to treat, requiring multiple operations with long hospital stays, prolonged antibiotic medication in high dosage, and extended periods of impairment. Still cure often cannot be obtained leading to amputation in many cases. The reasons for infection resistance against conventional antimicrobial therapies have been elucidated only in the last three decades, based on the pioneering work of William Costerton, showing that pathogens may change from the familiar planktonic forms into phenotypically different sessile forms after adhesion to unvascularized surfaces, forming the organized biocoenosis of a biofilm. Biofilm-embedded bacteria are present in all orthopedic infections and require much higher concentrations (up to 1,000 times) of antibiotics for elimination than their planktonic forms. For creating higher local antibiotic concentrations, carriers have been developed, but the frequently used devices made of poly(methyl methacrylate) cannot provide sufficient concentrations in the surrounding tissues and spaces and act as a substrate of biofilm colonization themselves. Antibiofilm substances have been investigated but are not yet available for clinical practice. Presently the only possibility of a biofilm-centered treatment is found in sophisticated techniques of debridement combined with increased antibiotic concentrations at the site of infection. Highly purified bone has been found to be an appropriate tool for storing and delivering the required amount of antibiotics in order to eliminate biofilm remnants after meticulous debridement. Additionally it offers the advantage of reconstructing osseous defects that always are present after surgically treated bone infection. Using antibiotic-impregnated bone graft treatment of infection, reconstruction and internal stabilization may be performed within a single operation. Long hospital

H. Winkler (✉)
Osteitis Centre, Privatklinik Döbling, Vienna, Austria
e-mail: heinz.winkler@speed.at

G.D. Ehrlich et al. (eds.), *Culture Negative Orthopedic Biofilm Infections*,
Springer Series on Biofilms 7, DOI 10.1007/978-3-642-29554-6_9,

stays and treatments associated with prolonged periods of pain and/or reduced mobility may be avoided.

## 1 The Problem of Bone Infection

Bacterial contamination of bone may occur via the bloodstream or through open wounds (e.g., fractures and ulcers). Orthopedic infections may occur as sequelae of traumatic episodes as well as during surgery, especially when foreign material is implanted, such as with total joint replacement or osteosynthesis. Damage to tissue leads to a decrease in blood supply and depression of immune responses, which can cause the formation of necrotic tissue and bacterial invasion. Bacteria, mostly staphylococci, are then able to bind to damaged tissue, replicate, and infection may ensue.

Unvascularized parts of bone may detach from the live bone and form a sequestrum. In some cases, the area of dead bone is too large to be resorbed; resultant changes in the periosteum, endosteum, and cortex cause the formation of new bone (involucrum) together with fibrous tissue. This mechanism is presumably designed to help isolate the infection.

Clinical signs persisting for longer than 10 days are associated with the development of necrotic bone and chronic osteomyelitis (COM). COM is characterized by the persistence of microorganisms, low-grade inflammation, and the presence of dead bone (sequestrum) and fistulous tracts. In most cases, there is poor local vascularization within a compromised soft-tissue envelope. The infected foci within the bone are surrounded by sclerotic, avascular bone covered by a thickened periosteum, scarred muscle, and subcutaneous tissue. Antibiotics and inflammatory cells cannot reach this avascular area, which is presumed to be the major reason that medical treatment of COM fails. COM generally cannot be eradicated without surgical treatment. The goal of surgery is to eliminate dead bone and achieve a viable vascularized environment.

Since the beginning of the twentieth century, clear recommendations for treatment of COM have been established (Editorial Comment 1919), already describing the principles of debridement (including devitalized scar tissue and sclerosis), dead space management, stabilization, and re-establishment of sufficiently vascularized tissue coverage. Diligent application of these principles may lead to favorable outcomes in the majority of cases. Even before the availability of antibiotics, these principles helped ensure that more than two-thirds of the patients could be treated successfully (Stephens 1921). It should be expected that the introduction of antibiotics should have provided for marked improvements in therapy of orthopedic infections. However, although antibiotics could lower the incidence of acute sepsis and fatal outcomes, the local healing of chronic cases was not influenced markedly (Reynolds and Zaepfel 1948; Rowling 1970).

Since it was thought that the poor penetration of the antibiotic into the infected bone site was the reason for lacking success, systemic application with high serum antibiotic concentrations was used for extended periods. These high serum levels were associated with nephrotoxicity, ototoxicity, and gastrointestinal side effects. Failure with systemic antibiotics led to consideration for local delivery of antibiotics directly to the infection site. Buchholz et al. were the first to mix antibiotics and poly(methyl methacrylate) PMMA for creating a local carrier (Buchholz and Engelbrecht 1970). Subsequently, Klemm et al. developed techniques using antibiotic-loaded bone cement formed into beads to be placed into debrided bone defects (Klemm 1979). Unfortunately, most of the antibiotic mixed into the beads did not get released. In fact, studies have shown that between 90 and 95% of the antibiotic remains trapped within the cement, leaving antibiotic bead chains ineffective as antimicrobial tool. While these beads might be considered to be a possible useful filler of dead space, cement is not biodegradable and therefore creates a physical barrier that prevents new bone from growing into the defect, requiring a second surgery to remove the beads.

Dead space management is another prerequisite for successful treatment of COM. Cavities caused by bacterial toxins, osteoclastic activity of immune defense, and surgical debridement need to be filled. The dead space must be managed to prevent recurrence; a significant bone loss might result in bone instability. Appropriate reconstruction of both the bone and soft-tissue defects may be needed. Several techniques have been successful when properly carried out (Prigge 1946), but, even with meticulous surgical technique, success cannot always be guaranteed.

One approach both for dead space management and reconstruction is bone grafting with primary or secondary closure using cancellous bone that can quickly become revascularized, with bone fragments becoming incorporated into the final bone structure. Bone taken from the patient's own body has been considered preferable; however, the availability of autologous bone is limited and harvesting leaves another surgical site that might add to overall surgical morbidity. Allogeneic bone (from a different human donor) has been used to overcome this drawback, with comparable outcome (Hazlett 1954). All grafting procedures still have a substantial failure rate due to resorption of the bone graft in the presence of persistent local infection. Recurrence of infection is found frequently and often requires several additional interventions.

## 2 Reasons for Failure

Most infection treatment failures can be explained by an understanding of differences in the efficacy of antimicrobial treatment against freely floating planktonic bacteria versus biofilm communities. In the 1970s, William Costerton began to elucidate the true reasons for infection resistance against conventional antimicrobial therapies, showing that pathogens may change from the familiar planktonic forms into phenotypically different sessile forms after adhesion to avital surfaces, forming the organized biocoenosis of a biofilm. Biofilm-embedded bacteria require much higher concentrations of antibiotics for elimination than their planktonic forms.

Enhanced resistance within biofilms may be attributed to several factors. Antimicrobial molecules must diffuse through the biofilm matrix in order to inactivate encased cells. The extracellular polymeric substances constituting this matrix present a diffusional barrier for these molecules by influencing either the rate of transport of the molecule to the biofilm interior or the reaction of the antimicrobial with the matrix material. Suci et al. (1994) demonstrated delayed penetration of ciprofloxacin into *Pseudomonas aeruginosa* biofilms. While only 40 s of exposure was required when using a sterile medium a biofilm-containing medium required 21 min.

Several bacteria respond to nutrient limitation and other environmental stresses by slowing bacterial growth. It appears that conditions that elicit decreased growth, such as nutrient limitation or the presence of toxic substances (e.g., antibiotics), favor the formation of biofilms. Biofilm-associated cells grow significantly more slowly than planktonic cells and, as a result, take up antimicrobial agents more slowly. For example, Anwar et al. (1992) found that older (10-day-old) chemostat-grown *P. aeruginosa* biofilms were significantly more resistant to tobramycin and piperacillin than were younger (2-day-old) biofilms. A dosage of 500 μg of piperacillin plus 5 μg of tobramycin per ml completely inactivated both planktonic and young (2-day-old) biofilm cells. Older (10-day-old) biofilm cell counts were reduced to only ~20 % by exposure to this dose. Similar results have been observed with several different combinations of bacteria and antimicrobial agents (Chuard et al. 1993; Desai et al. 1998; Amorena et al. 1999)[11–13].

Anthony Gristina showed that the same mechanisms responsible for antibiotic resistance in device-related infections likewise apply for bone infections (Gristina et al. 1985b). In those infections, our most obstinate opponents are not the familiar planktonic pathogens but their phenotypically different sessile forms embedded in an extracellular matrix, the glycocalyx (Gristina and Costerton 1985b; Costerton 2005). The surface of unvascularized bone and eventual implants acts as substratum for the attachment of bacteria and the formation of biofilms (Fig. 1a, b). Debridement may remove the vast majority bacteria, but even after a perfect debridement some colonies released from the biofilm during manipulation may find new habitats, able to colonize poorly vascularized or avital surfaces, and cause recurrence after an indefinite period of time. For this reason, simultaneous insertion of osteosynthetic material at the freshly debrided site is avoided and external fixators are encouraged for stabilization. Systemic antibiotic therapy and/or insertion of local antibiotic carriers may sterilize the site until delayed definitive supply becomes feasible. However, antibiotic levels reached by systemic antibiosis or local therapy with commercially available carriers cannot be effective in eliminating remaining biofilm clusters. For example, antibiotic bead chains are ineffective: 90% are covered with biofilms at removal (Neut et al. 2001). Most of the bacteria cultured from orthopedic implants show reduced susceptibility for antibiotics, even in their planktonic form (Tunney et al. 1998), whereas there is a significant correlation with previous use of gentamicin-loaded PMMA (Chang and Merritt 1992). All pathogens not identified with traditional cultures show elevated resistance against antibiotics (Zimmerli et al. 2004). Small colony variants (SCVs) require up to 100-fold antibiotic concentrations for elimination, but usually are

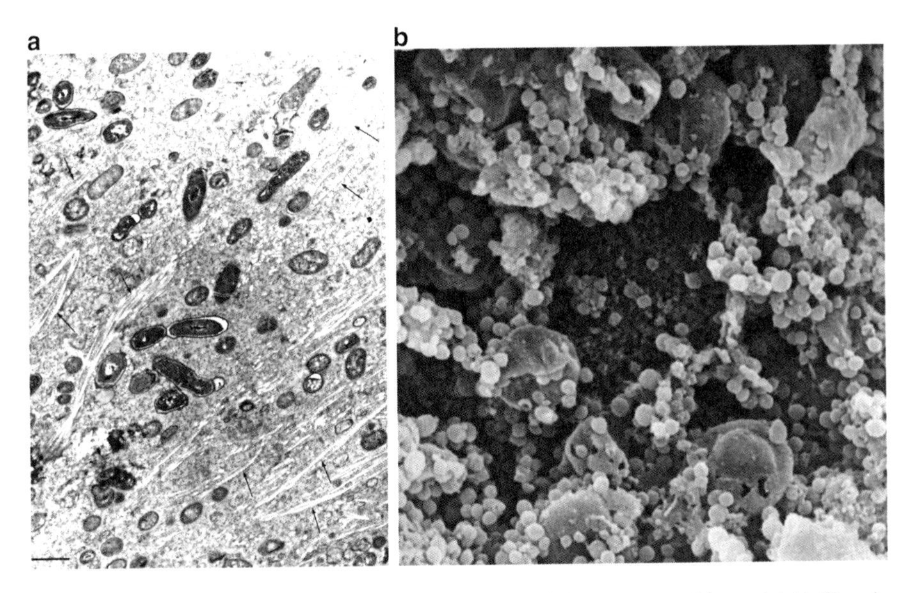


**Fig. 1** Direct demonstration, by electron microscopy, of the presence of bacterial biofilms in orthopedic infections. Transmission electron microscopy (**a**) and scanning electron microscopy (**b**) showing the presence of bacterial cells in bone infections that yielded negative cultures. In both micrographs the bacterial cells are arranged in biofilm clusters, and the *arrows* in (**a**) show that these microbial communities are integrated with the fibrous elements of muscle and tendon

accessible by systemic antibiosis, as long as the chosen antibiotics show that intracellular activity and application last long enough (von Eiff et al. 2006; Neut et al. 2007). Biofilm-embedded pathogens require up to 1,000-fold concentrations for elimination (Saginur et al. 2006) and are usually inaccessible for systemic antibiotic therapy as well as for antibiotics released from PMMA (van de Belt et al. 2001).

## 3 The Culture-Negative Orthopedic Infection

The gold standard for detection and classification of infection during the last 100 years has been bacterial culture. Most protocols for treating infected sites focus on microbiological results obtained perioperatively. However, it has turned out that the traditional and routinely used methods of culturing are likely to detect only a small proportion of the whole spectrum of pathogens possibly involved in orthopedic infection (Fux et al. 2003). Biofilm populations are often missed by conventional culture and only released planktonic forms may be detected. Sonication of explanted devices may dislodge and disrupt adherent biofilms and culturing sonication fluid is likely to raise the sensitivity of cultures significantly. For example, in patients who have received antimicrobial therapy within 14 days before culture, the sensitivities of periprosthetic tissue and sonicate-fluid culture rose from 45.0 to 75.0 % (Trampuz et al. 2007). Using immunofluorescence microscopy for visualizing dislodged pathogens after marking with specific antibodies reveals three times more colonies than seen with light microscopy. Amplification of bacterial genomes using PCR shows bacterial RNA in more than 70% of all revision hip replacement cases, including the so-called aseptic failures (Tunney et al. 1999; Nelson et al. 2005). The more sophisticated tools have confirmed that polymicrobial colonization is the rule rather than the exception after prolonged persistence of infection (Marrie and Costerton 1985). All of these findings indicate that the incidence and dimension of orthopedic infection is grossly underestimated by current culture detection methods (Tunney et al. 1999; Dempsey et al. 2007). As long as more sophisticated diagnostic tools are not commonly available, it seems advisable to expect infection with one or more pathogens even in cases where cultures reveal no growth. In the absence of more accurate predictive tools, clinical and radiological appearance should be primarily relied upon for clinical decision making.

## 4 Consequences for an Effective Therapy

An infected operative site cannot be sterilized by debridement alone. Debridement will remove most of bioburden, but even the most careful cleaning cannot prevent residual small bacterial colonies from being displaced to new habitats in niches of the debrided site. Antibiotic concentrations reached by systemic antibiosis or local therapy with established antibiotic carriers may provide eradication of planktonic residues but are not effective in eliminating microclusters disrupted from biofilms that may be the cause of recurrence after an indefinite period of time. Fragments of biofilms seem to be more vulnerable to antibiotics compared with intact biofilm systems (Fux et al. 2004; El-Azizi et al. 2005), but their elimination still requires concentrations exceeding those that might be achieved through systemic or conventional local antibiotic therapy. New local antibiotic carriers are necessary for eliminating residual biofilm fragments, providing sufficiently high local antibiotic concentrations for a prolonged period of time (Smith 2005).

After removal of infected implants and radical necrosectomy, bony defects always will be present. It is commonly accepted that whatever filler is used some kind of protection against recolonization with residual bacteria is needed. Dead space management after debridement may be performed with antibiotic-loaded cement, spacers, or bead chains. It should be kept in mind that those devices, besides their mechanical function, cannot be considered to be antimicrobial tools; their antibiotic content provides only short-lived prophylactic benefit against colonization with planktonic bacteria. These devices are not capable of sterilizing sites contaminated with sessile bacteria and provide no protection against biofilm colonization (Greene et al. 1998; Masri et al. 1998; Walenkamp 2001; Bertazzoni Minelli et al. 2004). Although most defects can be managed without structural support by allografts, e.g., by implanting a larger prosthesis reconstruction, even of small defects, is beneficial for success with possible further revisions. Allograft bone is widely used for reconstruction of bony defects and performs favorably in two-stage revisions of total joint replacement (Ammon and Stockley 2004). However, Unvascularized bone grafts are at risk of becoming contaminated and need protection.

## 5 Antibiotic Delivery

Since concentrations provided by systemic antibiotic therapy and commonly available carrier systems are insufficient for eliminating biofilm bacteria, new ways of antibiotic delivery are required. The criteria of antibiotics for efficacy against biofilms are different from those meant for action against planktonic bacteria. The required high antibiotic concentrations are only feasible by local application. Failure of antibiotics to cure prosthesis-related infection is not only due to poor penetration of drugs into biofilm but likely due to delayed antimicrobial effect on stationary bacteria in the biofilm environment.

When evaluating novel delivery systems, antibiotics must pass several qualifying tests. Few antibiotics have been identified to meet those criteria, with vancomycin being most widely evaluated. The majority of the pathogens involved in bone infection are Gram-positive and susceptible to vancomycin. Vancomycin shows an inferior tissue penetration compared with other antibiotics, which has been considered a disadvantage for systemic administration. Orally applied vancomycin is not resorbed. In local application, this apparent disadvantage turns into an advantage, since there is also very slow resorption and penetration from the bone into the vascular system. Vancomycin is one of the antibiotics with intracellular bactericidal activity and therefore should cover SCVs of staphylococci (Barcia-Macay et al. 2006). Furthermore, vancomycin is likely to penetrate glycocalyces very rapidly (Dunne et al. 1993; Darouiche et al. 1994; Jefferson et al. 2005). Once incorporated into biofilm, vancomycin displays strain-dependent bactericidal biofilm activity between 8 times (Rose and Poppens 2009) and 128 times (Gristina et al. 1989) the minimum inhibition concentration (MIC) of planktonic bacteria. Vancomycin shows superior bactericidal activity against biofilm-embedded staphylococci and especially MRSA compared with most other antibiotics (Smith et al. 2009). Keeping local vancomycin concentration at levels around 32 times the MIC of planktonic forms has been shown to reduce stationary phase pathogens by 2 logs within 24 h (Murillo et al. 2006). Vancomycin shows the least cytotoxic effect of all commonly used antibiotics (Edin et al. 1996) and is not likely to cause systemic side effects after local application (Buttaro et al. 2005a). It therefore may be suggested that local application of antibiotics with properties similar to vancomycin used with an appropriate delivery carrier may be a valuable tool against orthopedic infections. The carrier should provide for high initial levels to penetrate remaining biofilms rapidly and consequently keep the concentrations above the critical level (which, in the case of vancomycin, may be estimated to be between 200 and 500 mg/L) for a minimum of 72 h.

When loading bone grafts with antibiotics, the storage capability for antibiotics vastly exceeds those of PMMA (Witso et al. 1999, 2000; Buttaro et al. 2005b). Especially when using highly purified cancellous bone as a carrier, local concentrations of up to 20,000 mg/L can be released with vancomycin and up to 13,000 mg/L with tobramycin (Winkler et al. 2000). With this kind of impregnation, the whole amount of loaded antibiotic is available for antimicrobial activity and the activity remains far beyond the susceptibility of relevant pathogens for several weeks. These capacities make them more attractive for local therapy and allow using uncemented implants simultaneously. If cortical bone should become preferable, it can be loaded with antibiotics as well (Khoo et al. 2006). Using an adequate impregnation technique, antibiotics release may be similar to that seen with cancellous bone (Winkler et al. 2000). Kinetics is different but still capable of eliminating surrounding pathogens.

## 6 Why Use Purified Allograft Bone as a Carrier or Filler?

Besides the features that make bone an ideal carrier of antimicrobial substances (i.e., large surface, ideal binding capabilities), there are several biological issues that additionally contribute to persistent orthopedic infection. One of the predisposing factors for continuation of infection is inflammation. There are several reasons for inflammation, including mechanical (e.g., instability, foreign bodies), chemical, and immunological factors. When filling a previously infected site, techniques should be used to minimize the development of inflammation. PMMA or bone substitutes clearly are foreign bodies that tend to create an inflammatory response. Autologous bone (i.e., bone from the same donor) would be preferable but is often not available in sufficient amounts, requires additional surgery, and cannot be loaded sufficiently with antimicrobial substances. Unprocessed bone contains marrow consisting of cells and tissue likely to elicit an immunological reaction and fat eliciting inflammatory reactions. However, when removing all marrow and other tissue, the remaining material consists of a pure scaffold of collagen and enclosed minerals (Katz et al. 2009). These parts are identical within all species and, as such, are unlikely to cause an immune response (Friedlaender 1991). Highly purified bone matrix from the same species seems to show the highest biocompatibility of all available materials and additionally has advantages of similar load bearing capability and a favorable likelihood of incorporation into the host organism.

## 7 Choice of Implants

Whenever a new device is implanted into a recently infected site, the surgeon must be aware of increased risk of failure, with no significant difference between single- or two-stage procedures. Eventual removal, therefore, should be straightforward, with low risk of additional damage to the bony substance and an expectation of a good long-term result. This limits the choice of advisable implants. Cemented systems are less desirable since efficient cementing techniques will result in bonding with the underlying bone. Eventual removal, therefore, will be time consuming and possibly associated with further damage to the osseous structures. Furthermore, the addition of antibiotics reduces the biomechanical properties of cement (Klekamp et al. 1999; Baleani et al. 2008). Therefore, cemented revisions generally show inferior long-term results compared with uncemented techniques (Rothman and Cohn 1990; Lie et al. 2004). Bone cement (PMMA) has been shown to be the ideal substrate for bacterial attachment and replication of sessile bacterial phenotypes (Gristina et al. 1989). Addition of antibiotics to PMMA may be likely to act as a prophylactic aid against low bacterial numbers during the first days after implantation but cannot avoid colonization with high inocula (Gristina et al. 1985b), prevent biofilm formation on its surface (van de Belt et al. 2000; Dunne et al. 2007), or eliminate established biofilms (Tunney et al. 2007).

## 8 One Stage–Two Stage

There is ongoing discussion of how to treat infections related to orthopedic devices, especially in artificial joints. Most surgeons prefer first to remove the infected implants and later, in a second step, reimplant a new prosthesis. There is no consensus on the time interval between the procedures; mostly the second procedure is performed when clinical findings and laboratory parameters have returned to a normal appearance. In the meantime, systemic and/or local antibiosis can sterilize the infected site. However, the prolonged hospitalization and its associated costs, the delayed mobilization and rehabilitation, and the risk of additional surgery are marked drawbacks—especially in elderly patients. Is it worth to wait?

Few studies have compared one-stage and two-stage procedures (Jackson and Schmalzried 2000; Jamsen et al. 2009). Critical analysis, taking into account the total number of operations performed until clinical cure of infection, shows that there are no significant differences. This finding is not surprising when accepting the biofilm issue as the causal factor of recurrence, independent from surgical approach.

## 9 What Can We Do in Today's Practice?

Systemic application of antimicrobial substances seems to grant very limited chance of success in orthopedic surgery. In vitro data suggest some effect of rifampicin in combination with other antibiotics (Ghani and Soothill 1997; Gattringer et al. 2010) that partially could be verified in clinical practice (Konig et al. 2001). However, treatment requires several months and results are not predictable (Widmer et al. 1992), usually effective only in combination with surgical debridement (Zimmerli et al. 1998). This supports that incorporating methods to directly target biofilms may be the key to successful orthopedic procedures.

Several substances have been found that are likely to inhibit cell-to-cell communication inside biofilm formations ("quorum sensing"), making them more vulnerable to antimicrobial substances (Balaban et al. 2007). Ultrasound is another tool proven to disturb biofilm communities, helping to eradicate the enclosed pathogens (Pitt et al. 1994; Rediske et al. 1998; Carmen et al. 2005). In recent research, the possibilities of eradicating biofilm bacteria have focused more and more on local treatment. In clinical practice, the only option presently available is to provide sufficiently high concentrations of established antibiotics.

In order to eradicate microbial pathogens basic strategies need to be followed:

1. Localize microbial habitats as exactly as possible.
2. Drastically reduce microbe number and their means of livelihood by removing all identified avital material as radically as possible.
3. Disturb the living community of residual biofilm colonies by mechanically disrupting their established structures as thoroughly as possible.
4. Avoid re-establishment of colonization grounds by filling dead space with inaccessible material as completely as possible.
5. Eliminate sessile bacteria inside remaining fragments using antimicrobial substances in concentrations as high and as consistent as possible.

In addition, reconstruction of defects is favorable and often necessary to restore the function of the affected limb.

To address the problem of potentially undetected polymicrobial colonization, monotherapy should be reserved for cases with strong evidence of monomicrobial Gram-positive infection, i.e., acute onset of symptoms with typical clinical appearance (e.g., fever, pus) and unambiguous culture. Chronic infections as well as cases with prior infection-related surgery or inexplicit cultures should be considered polymicrobial and should be treated with a combination of two or more antibiotics. Combinations of vancomycin with tobramycin may be desirable, taking advantage of the synergistic activity of the two antibiotics (Watanakunakorn and Tisone 1982; Gonzalez Della Valle et al. 2001). This combined approach should likely cover most of the relevant pathogens since resistance to both antibiotics at the same time has not been found so far.

## 9.1 Clinical Recommendations

Based on the principles described above, a comprehensive protocol for the treatment of chronic bone infection has been established. Preoperatively, infected sites should be identified and mapped, using bone scans in combination with magnetic resonance imaging (MRI) and/or computed tomography (CT). All avital tissue

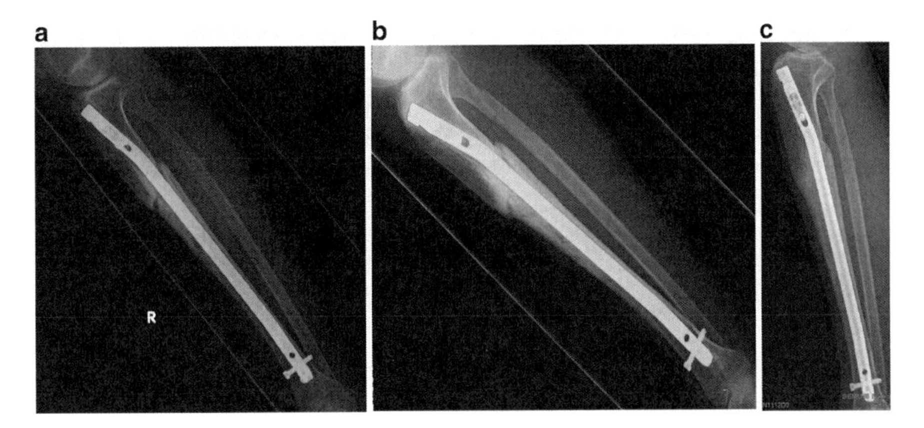


**Fig. 3** Radiographic follow-up of the revision in Fig. 2. (**a**) Immediately postoperative. (**b**) 5 months postop. (**c**) 10 months postop: fracture healed, graft incorporated, no sign of infection

osteomyelitis of long bones after trauma or hematogenous spread, diabetic foot syndrome, complicated corrective surgery, and infected endoprostheses, often including poor prognosis cases with pending amputation. The overall success rate, defined as an infection-free period of more than 2 years after one operation, was approximately 90%. Radiological incorporation of allografts occurred as is found after conventional bone grafting, with no adverse effects identified so far (Fig. 3).

## 10 Summary

Osteomyelitic lesions and infected implants may successfully be treated using thorough debridement in conjunction with antibiotic-loaded allograft bone, providing dead space management and reconstruction of deficient areas the same time. As long as the local antibiotic levels are higher than the dosage required for eliminating biofilm-embedded bacteria, contamination of alloplastic material is not anticipated. Internal fixation the same as re-implantation of endoprostheses may be performed simultaneously as performed under non-septic conditions. Using antibiotic-impregnated graft treatment of infection, reconstruction and internal stabilization may be performed within a single operation. Reinfection may occur in complex cases where secluded infection foci are not detected during debridement. An exact preoperative mapping of infected areas is therefore mandatory, along with postoperative follow up. Rerevisions are markedly less demanding when missed foci are detected early. Care should be taken for good soft tissue coverage, using muscle flaps in doubtful situations. Choice of stabilizing material should be considered carefully and the devices selected need to be left in situ for longer than usual.

Although the results of the new protocol adopting these principles seem very promising, we never can have the certainty of having cured bone infection.

Assuming that recurrence may occur within an unknown period of time, it should be the responsibility of the surgeon to provide a treatment reducing the burden for the patient to an absolute minimum. In this sense, it should be agreed that treatments should be kept as short and as pain-free as possible. Long hospital stays and treatments associated with prolonged periods of pain and/or reduced mobility should be avoided. The described protocol seems to be in conformance with these principles.

## References

Ammon P, Stockley I (2004) Allograft bone in two-stage revision of the hip for infection. Is it safe? J Bone Joint Surg Br 86:962–965

Amorena B, Gracia E, Monzón M et al (1999) Antibiotic susceptibility assay for Staphylococcus aureus in biofilms developed in vitro. J Antimicrob Chemother 44:43–55

Anwar H, Strap JL, Chen K, Costerton JW (1992) Dynamic interactions of biofilms of mucoid Pseudomonas aeruginosa with tobramycin and piperacillin. Antimicrob Agents Chemother 36:1208–1214

Balaban N, Cirioni O, Giacometti A et al (2007) Treatment of Staphylococcus aureus biofilm infection by the quorum-sensing inhibitor RIP. Antimicrob Agents Chemother 51:2226–2229

Baleani M, Persson C, Zolezzi C, Andollina A, Borrelli AM, Tigani D (2008) Biological and biomechanical effects of vancomycin and meropenem in acrylic bone cement. J Arthroplasty 23:1232–1238

Barcia-Macay M, Lemaire S, Mingeot-Leclercq MP, Tulkens PM, Van Bambeke F (2006) Evaluation of the extracellular and intracellular activities (human THP-1 macrophages) of telavancin versus vancomycin against methicillin-susceptible, methicillin-resistant, vancomycin-intermediate and vancomycin-resistant Staphylococcus aureus. J Antimicrob Chemother 58:1177–1184

Bertazzoni Minelli E, Benini A, Magnan B, Bartolozzi P (2004) Release of gentamicin and vancomycin from temporary human hip spacers in two-stage revision of infected arthroplasty. J Antimicrob Chemother 53:329–334

Buchholz HW, Engelbrecht H (1970) Depot effects of various antibiotics mixed with Palacos resins. Chirurg 41:511–515

Buttaro MA, Gimenez MI, Greco G, Barcan L, Piccaluga F (2005a) High active local levels of vancomycin without nephrotoxicity released from impacted bone allografts in 20 revision hip arthroplasties. Acta Orthop 76:336–340

Buttaro MA, Pusso R, Piccaluga F (2005b) Vancomycin-supplemented impacted bone allografts in infected hip arthroplasty. Two-stage revision results. J Bone Joint Surg Br 87:314–319

Carmen JC, Roeder BL, Nelson JL et al (2005) Treatment of biofilm infections on implants with low-frequency ultrasound and antibiotics. Am J Infect Control 33:78–82

Chang CC, Merritt K (1992) Microbial adherence on poly(methyl methacrylate) (PMMA) surfaces. J Biomed Mater Res 26:197–207

Chuard C, Vaudaux P, Waldvogel FA, Lew DP (1993) Susceptibility of Staphylococcus aureus growing on fibronectin-coated surfaces to bactericidal antibiotics. Antimicrob Agents Chemother 37:625–632

Costerton JW (2005) Biofilm theory can guide the treatment of device-related orthopaedic infections. Clin Orthop Relat Res (437):7–11

Darouiche RO, Dhir A, Miller AJ, Landon GC, Raad II, Musher DM (1994) Vancomycin penetration into biofilm covering infected prostheses and effect on bacteria. J Infect Dis 170:720–723

Dempsey KE, Riggio MP, Lennon A et al (2007) Identification of bacteria on the surface of clinically infected and non-infected prosthetic hip joints removed during revision arthroplasties by 16 S rRNA gene sequencing and by microbiological culture. Arthritis Res Ther 9:R46

Desai M, Bühler T, Weller PH, Brown MR (1998) Increasing resistance of planktonic and biofilm cultures of Burkholderia cepacia to ciprofloxacin and ceftazidime during exponential growth. J Antimicrob Chemother 42:153–160

Dunne WM Jr, Mason EO Jr, Kaplan SL (1993) Diffusion of rifampin and vancomycin through a Staphylococcus epidermidis biofilm. Antimicrob Agents Chemother 37:2522–2526

Dunne N, Hill J, McAfee P et al (2007) In vitro study of the efficacy of acrylic bone cement loaded with supplementary amounts of gentamicin: effect on mechanical properties, antibiotic release, and biofilm formation. Acta Orthop 78:774–785

Edin ML, Miclau T, Lester GE, Lindsey RW, Dahners LE (1996) Effect of cefazolin and vancomycin on osteoblasts in vitro. Clin Orthop Relat Res (333):245–251

Editorial Comment (1919) Treatment of chronic osteomyelitis of traumatic origin. Ann Surg 69:72–84

El-Azizi M, Rao S, Kanchanapoom T, Khardori N (2005) In vitro activity of vancomycin, quinupristin/dalfopristin, and linezolid against intact and disrupted biofilms of staphylococci. Ann Clin Microbiol Antimicrob 4:2

Friedlaender GE (1991) Bone allografts: the biological consequences of immunological events. J Bone Joint Surg Am 73:1119–1122

Fux CA, Stoodley P, Hall-Stoodley L, Costerton JW (2003) Bacterial biofilms: a diagnostic and therapeutic challenge. Expert Rev Anti Infect Ther 1:667–683

Fux CA, Wilson S, Stoodley P (2004) Detachment characteristics and oxacillin resistance of Staphyloccocus aureus biofilm emboli in an in vitro catheter infection model. J Bacteriol 186:4486–4491

Gattringer KB, Suchomel M, Eder M, Lassnigg AM, Graninger W, Presterl E (2010) Time-dependent effects of rifampicin on staphylococcal biofilms. Int J Artif Organs 33:621–626

Ghani M, Soothill JS (1997) Ceftazidime, gentamicin, and rifampicin, in combination, kill biofilms of mucoid Pseudomonas aeruginosa. Can J Microbiol 43:999–1004

Gonzalez Della Valle A, Bostrom M, Brause B, Harney C, Salvati EA (2001) Effective bactericidal activity of tobramycin and vancomycin eluted from acrylic bone cement. Acta Orthop Scand 72:237–240

Gorur ALD, Schaudinn C, Costerton JW (2009) Biofilm removal with a dental water jet. Compend Contin Educ Dent 30:1–6

Greene N, Holtom PD, Warren CA et al (1998) In vitro elution of tobramycin and vancomycin polymethylmethacrylate beads and spacers from Simplex and Palacos. Am J Orthop 27:201–205

Gristina AG, Costerton JW (1985) Bacterial adherence to biomaterials and tissue. The significance of its role in clinical sepsis. J Bone Joint Surg Am 67:264–273

Gristina AG, Oga M, Webb LX, Hobgood CD (1985) Adherent bacterial colonization in the pathogenesis of osteomyelitis. Science 228:990–993

Gristina AG, Jennings RA, Naylor PT, Myrvik QN, Webb LX (1989) Comparative in vitro antibiotic resistance of surface-colonizing coagulase-negative staphylococci. Antimicrob Agents Chemother 33:813–816

Hazlett JW (1954) The use of cancellous bone grafts in the treatment of subacute and chronic osteomyelitis. J Bone Joint Surg Br 36-B:584–590

Jackson WO, Schmalzried TP (2000) Limited role of direct exchange arthroplasty in the treatment of infected total hip replacements. Clin Orthop Relat Res (381):101–105

Jamsen E, Stogiannidis I, Malmivaara A, Pajamaki J, Puolakka T, Konttinen YT (2009) Outcome of prosthesis exchange for infected knee arthroplasty: the effect of treatment approach. Acta Orthop 80:67–77

Jefferson KK, Goldmann DA, Pier GB (2005) Use of confocal microscopy to analyze the rate of vancomycin penetration through Staphylococcus aureus biofilms. Antimicrob Agents Chemother 49:2467–2473

Katz J, Mukherjee N, Cobb RR, Bursac P, York-Ely A (2009) Incorporation and immunogenicity of cleaned bovine bone in a sheep model. J Biomater Appl 24:159–174
Khoo PPC, Michalak KA, Yates PJ, Megson SM, Day RE, Wood DJ (2006) Iontophoresis of antibiotics into segmental allografts. J Bone Joint Surg Br 88-B:1149–1157
Klekamp J, Dawson JM, Haas DW, DeBoer D, Christie M (1999) The use of vancomycin and tobramycin in acrylic bone cement: biomechanical effects and elution kinetics for use in joint arthroplasty. J Arthroplasty 14:339–346
Klemm K (1979) Gentamicin-PMMA-beads in treating bone and soft tissue infections (author's transl). Zentralbl Chir 104:934–942
Konig DP, Schierholz JM, Munnich U, Rutt J (2001) Treatment of staphylococcal implant infection with rifampicin-ciprofloxacin in stable implants. Arch Orthop Trauma Surg 121:297–299
Lie SA, Havelin LI, Furnes ON, Engesaeter LB, Vollset SE (2004) Failure rates for 4762 revision total hip arthroplasties in the Norwegian Arthroplasty Register. J Bone Joint Surg Br 86:504–509
Marrie TJ, Costerton JW (1985) Mode of growth of bacterial pathogens in chronic polymicrobial human osteomyelitis. J Clin Microbiol 22:924–933
Masri B, Duncan C, Beauchamp C (1998) Long-term elution of antibiotics from bone-cement: an in vivo study using the prosthesis of antibiotic-loaded acrylic cement (PROSTALAC) system. J Arthroplasty 13:331–338
Murillo O, Domenech A, Garcia A et al (2006) Efficacy of high doses of levofloxacin in experimental foreign-body infection by methicillin-susceptible Staphylococcus aureus. Antimicrob Agents Chemother 50:4011–4017
Nelson CL, McLaren AC, McLaren SG, Johnson JW, Smeltzer MS (2005) Is aseptic loosening truly aseptic? Clin Orthop Relat Res (437):25–30
Neut D, van De Belt H, Stokroos I, van Horn JR, van Der Mei HC, Busscher HJ (2001) Biomaterial-associated infection of gentamicin-loaded PMMA beads in orthopaedic revision surgery. J Antimicrob Chemother 47:885–891
Neut D, van der Mei HC, Bulstra SK, Busscher HJ (2007) The role of small-colony variants in failure to diagnose and treat biofilm infections in orthopedics. Acta Orthop 78:299–308
Pitt WG, McBride MO, Lunceford JK, Roper RJ, Sagers RD (1994) Ultrasonic enhancement of antibiotic action on gram-negative bacteria. Antimicrob Agents Chemother 38:2577–2582
Prigge EK (1946) The treatment of chornic osteomyelitis by the use of muscle transplant of iliac graft. J Bone Joint Surg Am 28:576–593
Rediske AM, Hymas WC, Wilkinson R, Pitt WG (1998) Ultrasonic enhancement of antibiotic action on several species of bacteria. J Gen Appl Microbiol 44:283–288
Reynolds FC, Zaepfel F (1948) Management of chornic osteomyelitis secondary to compound fractures. J Bone Joint Surg Am 30:331–338
Rose WE, Poppens PT (2009) Impact of biofilm on the in vitro activity of vancomycin alone and in combination with tigecycline and rifampicin against Staphylococcus aureus. J Antimicrob Chemother 63:485–488
Rothman RH, Cohn JC (1990) Cemented versus cementless total hip arthroplasty. A critical review. Clin Orthop Relat Res (254):153–169
Rowling DE (1970) Further experience in the management of chronic osteomyelitis. J Bone Joint Surg Br 52-B:302–307
Saginur R, Stdenis M, Ferris W et al (2006) Multiple combination bactericidal testing of staphylococcal biofilms from implant-associated infections. Antimicrob Agents Chemother 50:55–61
Smith AW (2005) Biofilms and antibiotic therapy: is there a role for combating bacterial resistance by the use of novel drug delivery systems? Adv Drug Deliv Rev 57:1539–1550
Smith K, Perez A, Ramage G, Gemmell CG, Lang S (2009) Comparison of biofilm-associated cell survival following in vitro exposure of meticillin-resistant Staphylococcus aureus biofilms to the antibiotics clindamycin, daptomycin, linezolid, tigecycline and vancomycin. Int J Antimicrob Agents 33:374–378

Stephens R (1921) Osteomyelitis following was injureis: based on the sutdy of 61 cases. J Bone Joint Surg Am 3:138–153
Suci PA, Mittelman MW, Yu FP, Geesey GG (1994) Investigation of ciprofloxacin penetration into Pseudomonas aeruginosa biofilms. Antimicrob Agents Chemother 38:2125–2133
Trampuz A, Piper KE, Jacobson MJ et al (2007) Sonication of removed hip and knee prostheses for diagnosis of infection. N Engl J Med 357:654–663
Tunney MM, Ramage G, Patrick S, Nixon JR, Murphy PG, Gorman SP (1998) Antimicrobial susceptibility of bacteria isolated from orthopedic implants following revision hip surgery. Antimicrob Agents Chemother 42:3002–3005
Tunney MM, Patrick S, Curran MD et al (1999) Detection of prosthetic hip infection at revision arthroplasty by immunofluorescence microscopy and PCR amplification of the bacterial 16 S rRNA gene. J Clin Microbiol 37:3281–3290
Tunney MM, Dunne N, Einarsson G, McDowell A, Kerr A, Patrick S (2007) Biofilm formation by bacteria isolated from retrieved failed prosthetic hip implants in an in vitro model of hip arthroplasty antibiotic prophylaxis. J Orthop Res 25:2–10
van de Belt H, Neut D, Schenk W, van Horn JR, van der Mei HC, Busscher HJ (2000) Gentamicin release from polymethylmethacrylate bone cements and Staphylococcus aureus biofilm formation. Acta Orthop Scand 71:625–629
van de Belt H, Neut D, Schenk W, van Horn JR, van der Mei HC, Busscher HJ (2001) Infection of orthopedic implants and the use of antibiotic-loaded bone cements. A review. Acta Orthop Scand 72:557–571
von Eiff C, Peters G, Becker K (2006) The small colony variant (SCV) concept – the role of staphylococcal SCVs in persistent infections. Injury 37(suppl 2):S26–S33
Walenkamp GH (2001) Gentamicin PMMA beads and other local antibiotic carriers in two-stage revision of total knee infection: a review. J Chemother 13(Spec No 1):66–72
Watanakunakorn C, Tisone JC (1982) Synergism between vancomycin and gentamicin or tobramycin for methicillin-susceptible and methicillin-resistant Staphylococcus aureus strains. Antimicrob Agents Chemother 22:903–905
Widmer AF, Gaechter A, Ochsner PE, Zimmerli W (1992) Antimicrobial treatment of orthopedic implant-related infections with rifampin combinations. Clin Infect Dis 14:1251–1253
Winkler H, Janata O, Berger C, Wein W, Georgopoulos A (2000) In vitro release of vancomycin and tobramycin from impregnated human and bovine bone grafts. J Antimicrob Chemother 46:423–428
Witso E, Persen L, Loseth K, Bergh K (1999) Adsorption and release of antibiotics from morselized cancellous bone. In vitro studies of 8 antibiotics. Acta Orthop Scand 70:298–304
Witso E, Persen L, Loseth K, Benum P, Bergh K (2000) Cancellous bone as an antibiotic carrier. Acta Orthop Scand 71:80–84
Zimmerli W, Widmer AF, Blatter M, Frei R, Ochsner PE (1998) Role of rifampin for treatment of orthopedic implant-related staphylococcal infections: a randomized controlled trial. Foreign-Body Infection (FBI) Study Group. JAMA 279:1537–1541
Zimmerli W, Trampuz A, Ochsner PE (2004) Prosthetic-joint infections. N Engl J Med 351:1645–1654

# Towards a New Paradigm in the Diagnosis and Treatment of Orthopedic Infections

**G.D. Ehrlich, J.W. Costerton, D. Altman, G. Altman, M. Palmer, C. Post, P. Stoodley, and P.J. DeMeo**

**Abstract** Imagine an office visit of an older, but still poised, oncologist with her orthopedic surgeon. The surgeon must somehow break the news that a prosthetic knee, installed some 10 months previously, must be removed in a two-stage procedure that will leave the lady immobile while an infection is brought under control, and a new knee can be implanted in 3 months. If this frightening specter was an ovarian cancer, the oncologist would ask if the genome of her cancer had been sequenced and whether estrogen receptors were present on her malignant cells, so that a rational treatment strategy could be devised. The surgeon would say that the aspirate he took on the previous visit had grown "Staphylococcus epidermidis," in small numbers that might be contaminants from the technician's hands, and that most strains of *Staphylococcus epidermidis* these days were resistant to methicillin so she had better go on an aminoglycoside until her surgery early next week. And the twenty-first century would have ground to a sickening halt, on the residue of nineteenth century techniques, right there in the surgeons' office.

G.D. Ehrlich (✉)
Center for Genomic Sciences, Allegheny-Singer Research Institute, Pittsburgh, PA, USA
e-mail: gehrlich@wpahs.org

J.W. Costerton • D. Altman • G. Altman • M. Palmer • P.J. DeMeo
Department of Orthopedic Surgery, Allegheny General Hospital, Pittsburgh, PA, USA

C. Post
Allegheny-Singer Research Institute, Pittsburgh, PA, USA

P. Stoodley
Faculty of Engineering, University of Southampton, Southampton, UK

G.D. Ehrlich et al. (eds.), *Culture Negative Orthopedic Biofilm Infections*,
Springer Series on Biofilms 7, DOI 10.1007/978-3-642-29554-6_10,

# 1 Introduction

Culture techniques have persisted, virtually unchanged, from the middle of the 1870s (Koch 1884). The methods currently used to ascertain antibiotic sensitivity date from the middle of the 1950s (Bauer et al. 1966). These arcane technologies constitute the entirety of the basic FDA-approved bacterial diagnostic facilities enshrined in the clinical microbiology laboratories of all of our hospitals, throughout the developed world. Modern PCR-based methods have crept into occasional use in the detection and identification of specific pathogenic bacteria, but their coverage is spotty and limited to diseases in which cultures are obviously too slow or insensitive to be useful. Worse, the whole nonmedical microbiological /adopted DNA-based molecular methods for bacterial population analysis in the mid-1970s, and even the NIH-sponsored project to study the "microbiome" of healthy humans depends completely on this modern technology. The modern world has left cultures behind, as a means of population analysis and bacterial detection, but these methods persist in our hospitals and form the only current microbiological basis on which treatment for bacterial infections is predicated.

The only fair way to compare new techniques with their older counterparts is to compare them in a real clinical setting, and to record their performances on a "level playing field." This side-by-side comparison has been made in several medical fields, notably in middle-ear infections and in chronic wounds (chapter "Culture-Negative Infections in Orthopedic Surgery"), but we will begin by recording such a comparison in orthopedic surgery.

# 2 Comparison of Cultures with Molecular Methods in Orthopedic Surgery

Orthopedic surgery offers an opportunity to compare and contrast cultures with DNA-based molecular techniques for the detection and identification of bacteria, in tissues that are normally sterile, and in infections that typically involve only a limited number of pathogenic species. Bacterial incursions into primary joint replacements might be expected to involve a small number of contaminating species, and non-unions of closed fractures would be expected to be similarly lacking in microbial diversity. Open fractures would, of course, be subject to contamination from a wider variety of bacteria from the environment in which the trauma occurred. Previous studies of device-related and other chronic orthopedic infections have shown, unequivocally, that the causative organisms grow in well-developed biofilms (Fig. 1), so that all of the difficulties in detecting these sessile community-based organisms should be anticipated in this study.

The cumulative effect of these assumptions is that we resolved to compare culture methods, as conducted in an accredited and well-regarded clinical laboratory, with DNA- and RNA-based molecular methods for the detection and identification of biofilm bacteria in orthopedic infections.

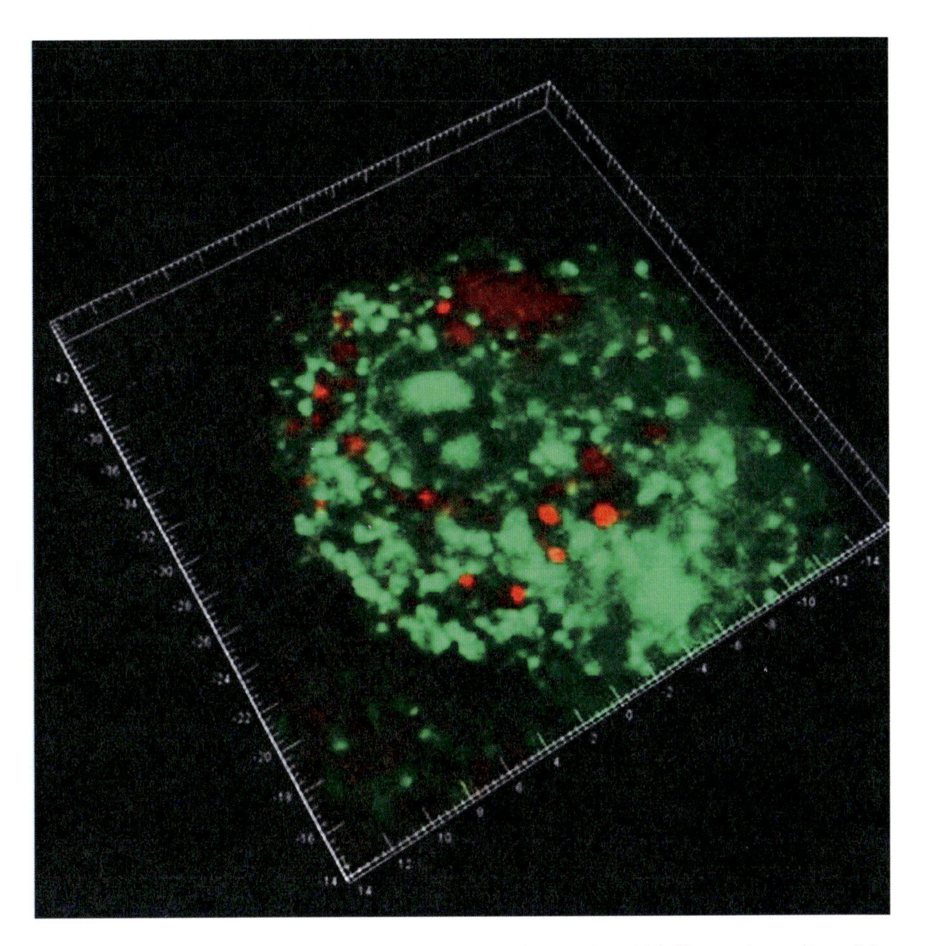

**Fig. 1** Confocal micrograph of bacteria growing in a slime-enclosed biofilm on the surface of the plastic component of an infected ankle joint. The bacterial cells (*Staphylococcus aureus*) have been stained with the Molecular Probes live/dead kit, which shows living cells as *green spheres*, and dead cells and host tissues as *red structures*. The whole community of bacteria is enclosed in an amorphous matrix, which stains *pale green* in this projection, and this represents the bacterial exopolysaccharide and extruded bacterial DNA with adsorbed host materials that make up the extracellular matrix & the biofilm. The numerals on the edges of the frame indicate the scale in microns. From (Stoodley et al. 2011)

## 2.1 *Preliminary Data from a Prospective Double Blind Study*

The DNA-based technology we chose for the detection and identification of bacteria in this definitive comparison of culture with molecular methods is the multiplex Ibis system, which uses the mass of PCR-generated amplicons (Ecker et al. 2008) to match base ratios to a database containing thousands of known bacterial species. This system has the potential advantages of detecting and identifying all of the prokaryotic organisms present in the sample, of detecting specific genes that confer antibiotic resistance, and of accomplishing these molecular feats in <6 h. The RNA-based technology used as a gold standard in this study is the fluorescence *in situ* hybridization (FISH) technique that has been used to identify individual cells, at the species level, and to "map" biofilm populations in relation to the living and inert surfaces they colonize in virtually all ecosystems (Nistico et al. 2009).

We have begun a prospective double blind study, in which we compare the sensitivity and specificity of routine cultures with DNA- and RNA-based molecular methods, in patients with non-unions following osseous fractures, and Table 1 presents some preliminary data from this study. In this study we submit intra-operative specimens to an accredited hospital-based microbiology laboratory that does not use any special techniques (e.g., sonication) to recover biofilm bacteria, does not incubate any specimens anaerobically, and discards all cultures at 5 days. We have chosen this *modus* because we realize that most departments of Orthopedic Surgery rely on similar labs for their supportive microbiology, and that sophisticated culture facilities like those at the Mayo Clinic and in Lausanne and Geneva are a rarity. Because neither the Ibis platform nor the FISH technique is FDA-approved as a basis for therapeutic strategies, we have simply collected data on the same specimens submitted for culture, to gain an impression of the impact that molecular data may have when the Ibis is marketed in the future.

We will not attempt any statistical analysis of our data, because the numbers are too small, but it may be useful to note trends that may emerge as solid data when the numbers in this study reach their planned total of >100 patients. Because molecular methods are often dismissed as being hypersensitive, it is important to note that 3/20 patients (patients18–20) showed negative data from cultures and were negative by Ibis (in both specimens) and negative by the FISH technique. A proportion of the patients who chose surgery in connection with fractures do so to have hardware removed for convenience and, when the code on the study is broken, we will determine if these "negative controls" correspond to the negative specimens. Because the new molecular techniques will be much faster than cultures (±6 h), and because pre-surgical aspirates from overtly infected non-unions and prosthetic joints may be useful in designing antibiotic coverage for corrective surgery, it is interesting to note that 3/20 patients (patients 1–3) in this series had their Staphylococcal cultures confirmed by one or more Ibis results, and by FISH probe analysis of their tissues. These data, involving two cases with *S. aureus* and one case with *S. epidermidis*, are reassuring because the greatest initial benefit of the molecular techniques may be their ability to provide positive data to support rapid treatment and their ability to provide negative data that can facilitate the release of patients with suspected infections without having to wait 5 days for culture results. Ibis data do not always agree with culture data (patients 4 and 5), and infections may indeed be polymicrobial, as the FISH data indicate that both nucleic acid-based methods show that bacteria are present, and that the debridement undertaken during this surgery removes the dead tissues and the biofilms that have combined to negatively affect the healing of the fracture.

**Table 1** Comparison of cultures with the Ibis method, and with direct bacterial visualization by the FISH technique, in a cohort of 20 patients undergoing surgery for non-union of fractures of long bones

| | Culture | Ibis—tissue | Ibis—membrane | FISH and live/dead |
|---|---|---|---|---|
| Culture, Ibis, and FISH agree | | | | |
| 1 | *S. aureus* | *S. aureus* | *S. aureus* | *S. aureus* biofilms in tissue |
| 2 | *S. aureus* | *S. aureus* | *S. aureus* + five other species | *S. aureus* in biofilms in tissue |
| 3 | *S. epidermidis* | *S. epidermidis* | No membrane | Staph. species in biofilms |
| Culture and Ibis disagree. FISH confirms bacterial presence | | | | |
| 4 | MRSA | *Burkholderia cenocepacia* | *Enterococcus faecalis* | Universal probe = bacteria in biofilms |
| 5 | MRSA | *S. epidermidis* | *Pseudomonas nitroreducens* | Universal probe = bacteria in tissue |
| Culture negative, Ibis positive, FISH confirms Ibis at species level | | | | |
| 6 | Negative | *Pseudomonas aeruginosa* | No membrane | *Pseudomonas aeruginosa* biofilms |
| 7 | Negative | *Pseudomonas nitroreducens* | *Streptococcus pyogenes* | Pseudomonas and Streptococcal biofilms |
| 8 | Negative | *Staphylococcus capitis* | No membrane | Staphylococcal sp. biofilm on screw |
| 9 | Negative | *Enterococcus faecalis* | No membrane | *E. faecalis* in tissue |
| Culture negative, Ibis positive, FISH confirms bacterial presence | | | | |
| 10 | Negative | *Actinobacillus capsulatus* | No membrane | Universal probe = extensive biofilm |
| 11 | Negative | *E. faecalis* + Cladosporium sp. | No membrane | Universal probe = biofilms |
| 12 | Negative | *Enterococcus faecalis* | Corynebacterium sp. | Universal probe = bacterial biofilms |
| 13 | Negative | *Enterococcus ascini* | Bordetella sp. | Live/dead = huge complex biofilms |
| 14 | Negative | Negative | Acinetobacter sp. + *S. aureus* | Universal probe = many integrated cells |
| 15 | Negative | *Campylobacter concisus* | Negative | Universal probe = bacteria in biofilms |
| 16 | Negative | Negative | Lactobacillus sp. | Universal probe = compact biofilm |
| Culture negative, Ibis positive, FISH negative | | | | |
| 17 | Negative | *Brevundimonas bacteroides* | *T. denticola* + *E. faecalis* | Negative |

(continued)

**Table 1** (continued)

| | Culture | Ibis—tissue | Ibis—membrane | FISH and live/dead |
|---|---|---|---|---|
| Culture negative, Ibis negative, FISH negative | | | | |
| 18 | Negative | Negative | Negative | Negative |
| 19 | Negative | Negative | Negative | Negative |
| 20 | Negative | Negative | Negative | Negative |

Comparison of the data from aliquots of the same intra-operative samples that have been divided for culture, Ibis analysis, and the imaging of bacterial cells by the FISH technique. The data labeled "Ibis membrane" was specially collected, when it was present in the surgical field, and it is the amorphous material that is sometimes present at a fracture site which looks (and smells) different from normal tissues and may represent the actual biofilm that complicates healing in some cases. Cultures were reported as positive if any of several specimens yielded growth on plating, before the plates were discarded on day 5. FISH probes were chosen on the basis of the culture and Ibis data, and some were specific at the species or genus level, while the universal probe was the Eubac 338 construct that reacts with the 16 S rRNA of all Eubacteria. The live/dead probe used in one instance was the BacLight kit purchased from Molecular Probes, Eugene, Oregon.

The next group of patients (patients 6–9) are of particular interest in this preliminary assessment of data from this cohort because cultures were negative, but one or more specimens showed the presence of recognized orthopedic pathogens using the Ibis methods, and the FISH technique confirmed the actual presence of bacteria of that species/genus in the affected tissues. Patient 6 was culture-negative, but the Ibis detected *P. aeruginosa* in the tissue near the fracture site, and the FISH preparation proved that cells of this organism were present and well integrated in the tissues, in a manner that indicates invasion rather than contamination. Patient 7 was culture-negative, but the Ibis detected a *Pseudomonas* species in the tissues and *Streptococcus pyogenes* in the pseudo-membrane, and FISH probing with both the *Pseudomonas* genus probe and the species-specific probe for *S. pyogenes* showed very extensive mixed species biofilms. Patient 8 was culture-negative, but the Ibis detected *Staphylococcus capitis* [a coagulase-negative staphylococcal species (CONS)] in the tissue near the fracture site, and FISH probing with a staphylococcal genus probe clearly showed the presence of large numbers of coccoid bacteria on the surface of a screw from the internal fixation (Fig. 2b). Patient 9 was culture-negative, but the Ibis detected *Enterococcus faecalis* in the tissue near the fracture site, and the use of a species-specific *E. faecalis* FISH probe showed the presence of large numbers of coccoid cells fully integrated into the tissues in a manner that can only indicate invasion and proliferation of pathogenic bacteria (Fig. 2a).

These data are both disturbing and unequivocal. Cultures failed to detect two species of *Pseudomonas* a CONS and *E. faecalis*, all of which are recognized pathogens in the orthopedic setting, and direct microscopy of the affected tissues shows the presence of large numbers of the invading bacteria in the tissues. FISH probes detect bacterial 16 S rRNA, which is normally absent from musculoskeletal tissues, and the reactive structures (Fig. 2) are bacterial cells $<1$ μm in size which cannot be confused with any human cell structures, so the evidence is airtight. In these four cases, culture failed to detect the pathogens, Ibis succeeded in detecting and identifying the pathogens, and FISH confirmed that the molecular technique yielded correct data.

In the largest cohort of patients (10–16) in Table 1 cultures were negative, but Ibis was positive on one or more samples, but we were only able to confirm the presence of bacteria (and not the species identity) because we lacked the FISH probes necessary to provide species or genus identification. When the study code is broken, and we know full clinical details, more revelations may be forthcoming, but at least some of these cases may represent biofilm infections that have escaped detection by culture techniques. Patient 10 is of special interest because Ibis detected *Actinobacillus capsulatus* in the tissue sample, at a level (2,741 genomes/well) higher than any other bacterium in this series, and the universal FISH probe found huge biofilms that occupied large areas of the affected tissue. This patient had an open fracture of the tibia (Grade 2) and this organism may have gained access to the fracture site from the environment, via the broken skin, thus it must now be added to the lexicon of orthopedic pathogens because it was associated with a non-union that lasted at least 3 months. Patients 11–13 were culture negative, but Ibis detected *E. faecalis* or *Enterococcus ascini* in one or more intra-operative

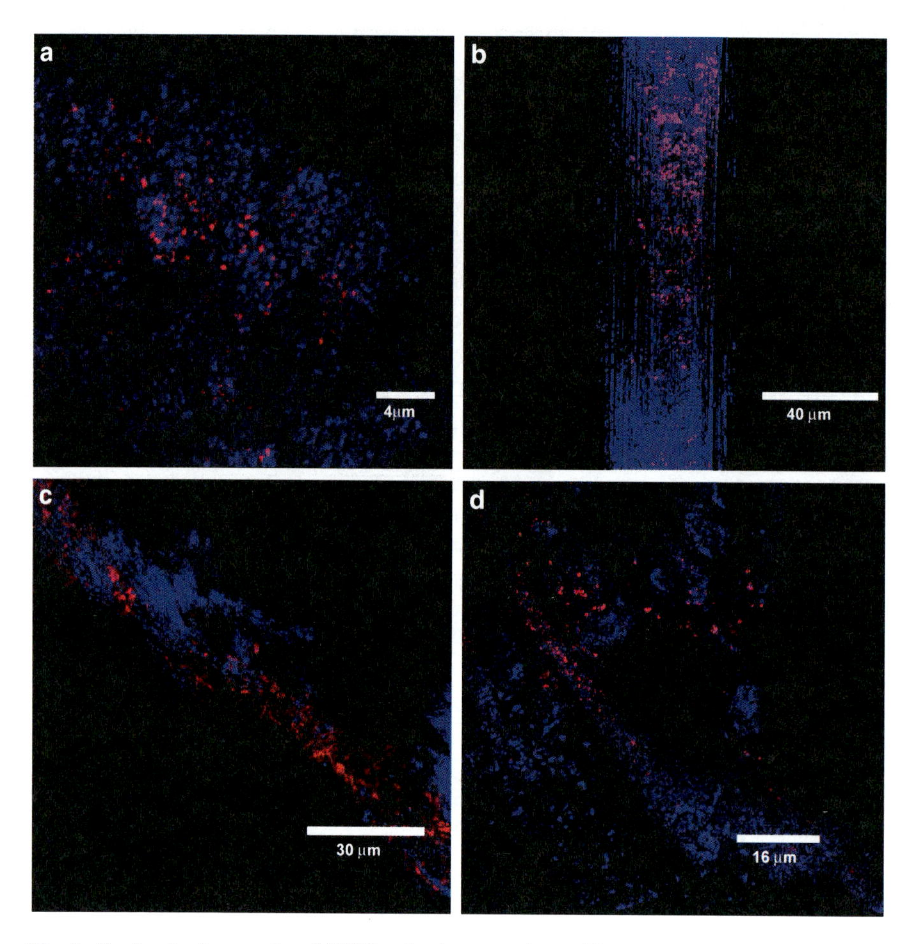


**Fig. 2** Confocal micrographs of FISH-stained preparations of intra-operative samples of tissue and hardware from non-unions of long bone fractures. Confocal micrographs of tissues and hardware from non-union cases that consistently yielded negative culture data, showing the presence of large numbers of biofilm bacteria by their unequivocal reactions with species-specific and universal FISH probes. (**a**) Tissue from a culture-negative patient in which the Ibis detected *Enterococcus faecalis*, and in which the species-specific FISH probe shows the presence of very large numbers of cells of this organism. (**b**) A screw from a culture-negative patient in which the Ibis detected *Staphylococcus capitis*, and in which the staphylococcal genus-specific FISH probe shows the presence of a coherent biofilm formed by this organism on this inert surface. (**c**) Tissue from a culture-negative patient (15) in which Ibis detected *Campylobacter concisus*, and in which the use of the universal Eubac 338 probe shows the presence of very large numbers of bacteria (species unknown) in well-developed biofilms that occupy a large proportion of the tissue. (**d**) Tissue from a culture-negative patient (12) in which the Ibis detected *Enterococcus faecalis* and Corynebacterium sp. and in which the use of the universal Eubac 338 probe shows the presence of large numbers of bacteria in well-developed biofilms

samples, and the universal FISH probe showed the presence of coccoid bacterial cells in the affected tissues (Fig. 2d) in patient 12. These enterococcal species are recognized orthopedic pathogens, and the FISH probes show that they have invaded and occupied large areas of tissue, so the non-unions they have caused can be attributed to a bacterial biofilm presence that has escaped detection by conventional culture techniques. Patients 14–16 represent an enigmatic cohort. *S. aureus* was detected in very low numbers (8 genomes/well) in patient 14, the *Campylobacter concisus* detected by Ibis in patient 15 may not be a pathogen, and the Lactobacillus sp. detected by Ibis in patient 16 was both in low numbers (6 genomes/well) and of dubious pathogenic potential. However, the universal FISH probe showed that extensive bacterial biofilms were present in all cases, and non-unions may have resulted from their presence, and an examination of Fig. 2c shows that the cells of *C. concisus* had formed very invasive biofilms in the tissue of patient 15. Patient 17 may be an anomaly because, while cultures were negative and Ibis showed the presence of large numbers of three bacterial species, extensive searches of two separate FISH preparations failed to show any bacterial cells. This may be simply a function of the spatial heterogeneity of polymicrobial infections (Hall-Stoodley et al 2006).

# 3 The Way Forward

The art of diagnosis is complex and exacting, and this heavy responsibility lies entirely in the province of physicians and surgeons who must weigh data from many sources to arrive at correct conclusions to enhance the welfare of their patients. The preliminary data from this prospective double blind study of non-unions, comparing cultures with DNA-based and RNA-based molecular methods, indicate that laboratories may soon be able to provide clinicians with much better microbiological data on which to base their diagnoses of orthopedic infections. The difficulties posed by the biofilm mode-of-growth of the bacteria involved resulted in positive cultures from only 5/20 patients in this non-union series (patients 1–5), and the cultures from patients 1–3 were confirmed by the independent Ibis and FISH techniques. The fact that these infections involved *S. aureus* and *S. epidermidis* reinforces the general impression that these well-known pathogens grow readily on routine media and yield colonies in <5 days. If small bacterial numbers and dubious pathogenicity are excluded (patients 14–17), the Ibis technology detected bacteria in 13/20 of the same cohort of patients, and it is clear that many of these organisms are recognized pathogens (e.g., *E. faecalis*) whose presence as biofilms would be expected to delay the healing of fractures. Because the Ibis technology does not require bacterial growth, its detection of bacteria is based entirely on their content of DNA (Ecker et al. 2008) and not on the choice of media or growth conditions, so it detects and identifies bacteria in many more cases of non-union. The detection of *Actinobacillus capsulatus*, which is a recognized equine pathogen (Blackall et al. 1997), in very large numbers and in biofilms near a non-union secondary to an open fracture in patient 10 sets the stage for a diagnostic drama that would certainly challenge Dr. House and might even bring fame to an obscure orthopedic surgeon. More somberly, microbiology can serve all branches of medicine more efficiently if it is allowed to use its best modern techniques, the ones it uses in its own internal research, to provide clinicians with meaningful and comprehensive data on which to base the diagnoses that affect the lives of the millions of patients (Wolcott et al. 2010) currently affected by biofilm infections.

These preliminary orthopedic data indicate that new molecular methods (e.g., Ibis) will yield microbiological data that are more useful than that provided by traditional cultures, in supporting the diagnosis of infection in musculoskeletal tissues that are normally sterile. If we then seize on a distillation of the salient points, we will conclude that the detection of biofilm bacteria by the measurement of one of their components (e.g., DNA) is vastly preferable to their detection by persuading a few planktonic cells shed from the biofilm community to grow and produce colonies on the surfaces of agar plates. To take a particularly grotesque example, we monitor the colonization of vascular hardware by removing catheters cutting off the tips and rolling them on the surface of an agar plate, in the time-honored "Maki" technique (Maki et al. 1977). In this exercise we depend on the fact that biofilms on the external surface of the catheter usually shed a small number of planktonic cells, which will detach from the plastic if the technician hits exactly the correct rolling rhythm, and will give rise to visible colonies if the conditions on the agar surface happen to be optimal. In the new molecular mind-set, we would simply remove blood (or locking solution) without removing the catheter, detect and identify bacteria by analyzing the DNA they shed continuously into their fluid environment, and support clinical decisions with species identity and antibiotic sensitivity data.

# 4 Valete

When molecular methods for the detection and identification of bacteria are accepted in one of the medical silos (e.g., orthopedics), the common problem of the diagnosis of biofilm infections will prompt a similar conversion in other specialties. Our dependence on cultures to detect occasional planktonic cells released from biofilms fuels speculation that many chronic diseases are caused by bacterial biofilms (Costerton et al. 1999). We are already involved in a large multicenter NIH-funded project to detect bacteria in urine from male patients with suspected prostatitis and female patients with chronic urinary tract infections, and Ibis data will establish the presence and identity of biofilm bacteria that can then be located in infected tissues by the FISH technique. When we reach the tipping point in our creeping realization that biofilm infections predominate in modern medicine (Costerton et al. 1999), and that these biofilm infections are very difficult to detect using culture methods (Veeh et al. 2003), all of the medical silos will adopt molecular methods to support the diagnosis of infection. When the species identity and antibiotic sensitivity of the bacterial species responsible for these chronic biofilm infections are known, direct microscopy with FISH probes will locate the biofilms in the affected tissues, and the live/dead probe will enable us to monitor the effects of anti-biofilm signals (Balaban et al. 2005) and novel pharmaceuticals. When all of these perceptions are realized, microbiology will be truly supportive of medicine, by using its best and most advanced technologies to provide accurate data to enable the accurate diagnosis and effective treatment of bacterial biofilm infections.

## References

Balaban N, Stoodley P, Fux CA, Wilson S, Costerton JW, Dell'acqua G (2005) Prevention of Staphylococcal biofilm-associated infections by the quorum sensing inhibitor RIP. Clin Orthop Relat Res 437:48–54

Bauer AW, Kirby WMM, Sherris JC, Turck M (1966) Antibiotic susceptibility testing by a standardized single disc method. Am J Clin Pathol 45:493–496

Blackall PJ, Bisgaard M, McKenzie RA (1997) Characterisation of Australian isolates of Actinobacillus capsulatus, Actinobacillus equuli, Pasteuraella caballi and Bisgaard taxa 9 and 11. Aust Vet J 75:52–55

Costerton JW, Stewart PS, Greenberg EP (1999) Bacterial biofilms: a common cause of persistent infections. Science 284:1318–1322

Hall-Stoodley L, Hu FZ, Gieseke A, Nistico L, Nguyen D, Hayes J, Forbes M, Greenberg DP, Dice B, Burrows A, Wackym PA, Stoodley P, Post JC, Ehrlich GD, Kerschner JE (2006) Direct detection of bacterial biofilms on the middle-ear mucosa of children with chronic otitis media. JAMA 296:202–211.

Ecker DJ, Sampath R, Massire C, Blyn LB, Hall TA, Eshoo MW, Hofstadler SA (2008) Ibis T5000: a universal biosensor approach for microbiology. Nat Rev Microbiol 6:553–558

Koch R (1884) Die aetiologie der tuberkulose, mittheilungen aus dem kaiserlichen. Gesundhdeitsamte 2:1–88

Maki DK, Weise CE, Sarafin HW (1977) A semi-quantitative method of identifying intravenous catheter-related infections. N Engl J Med 296:1305–1309

Nistico L, Gieseke A, Stoodley P, Hall-Stoodley L, Kerschner JE, Ehrlich GD (2009) Fluorescence in situ hybridization for the detection of biofilms in the middle ear and upper respiratory tract mucosa. Methods Mol Biol 493:191–213

Stoodley P, Conti SF, DeMeo PJ, Nistico L, Melton-Kreft R, Johnson S, Darabi A, Ehrlich GD, Costerton JW, Kathju S (2011) Characterization of a mixed MRSA/MRSE biofilm in an explanted total ankle arthroplasty. FEMS Immunol Med Microbiol 62:66–74

Veeh RH, Shirtliff ME, Petik JR, Flood JA, Davis CC, Seymour JL, Hansmann MA, Kerr KM, Pasmore ME, Costerton JW (2003) Detection of Staphylococcus aureus biofilm on tampons and menses components. J Infect Dis 188:519–530

Wolcott RD, Rhoads DD, Bennett ME, Wolcott BM, Gogokhia L, Costerton JW, Dowd SE (2010) Chronic wounds and the medical biofilm paradigm. J Wound Care 19:45–54

# Index



G.D. Ehrlich et al. (eds.), *Culture Negative Orthopedic Biofilm Infections*, Springer Series on Biofilms 7, DOI 10.1007/978-3-642-29554-6,

Printed by Publishers' Graphics LLC
BT20121223.19.21.16